# THE CHESAPEAKE CONSPIRACIES

*A LEGAL THRILLER*

**Nino Lama**

## Disclaimer

*This is a work of fiction. Names, characters, businesses, places, events, and incidents are either the products of the author's imagination or used in a fictitious manner. Any resemblance to actual persons, living or dead, or actual events is purely coincidental.*

# Contents

# PROLOGUE

## 1982

The union between Ronald J. Bryce and his wife, Emily, had always teetered on the brink. They had exchanged vows just as Ronald's honorable discharge from the U.S. Army coincided with the end of the Vietnam war in 1975. The horrors he had witnessed in the unforgiving jungles of Vietnam had etched their mark on his psyche, erupting in relentless life-numbing flash-backs and black tar swamp-like nightmares that left him drenched in sweat, trembling in the dead of night. Upon Ron's return from Vietnam, there were no sympathetic ears to diagnose the haunting specter of PTSD. Those were different times, and the remedy of choice was drowning one's demons in alcohol. Ron, Sr., steadily descended into a pit of despair, self-medicating with whiskey as he grappled with the ghosts of war.

Their home was a modest raised-ranch on Maryland's eastern shore, a home financed through the generous provisions of the GI Bill—no down payment, negligible interest rate. Ron eked out a living as a long-haul trucker for TransNation Haulers, hauling boiling asphalt across the nation for highway construction projects. He was rarely home, spending just a handful of days each month under his own roof with his family.

Emily hadn't always been the thin and frail woman she was now, worn down by her husband's drastic mood swings, sleepless nights and rages. She did all she could to keep peace in the house, food on the table, and a smile for her husband when she could manage one.

Ronald Bryce Jr., affectionately known as Ronnie, arrived in the world in 1976, a beacon of hope, breathing new life into their strained marriage. At six years old, he was the guiding light in his father's otherwise shadowy world, though their encounters were sporadic at best. Though he clung tightly to his mother's apron strings, young Ronnie held a deep longing for the rare weekends spent with his father.

***

It was on a long-haul drive to deliver a load of asphalt for the construction of the Bay Bridge across the Chesapeake Bay that he met an avid—almost fanatic—boater at a roadhouse bar in Annapolis. After a few rounds of Jim Beam, Ron Sr. was convinced a boat was what he and his family needed to find some semblance of family and for him to find some happiness, and maybe even peace, unite the family.

Emily couldn't have been happier to hear Ron's plan. She knew they needed *something* new in their lives, and she was sure Ronnie Jr. would love it. And so, they borrowed money, bought a modest, used wooden twenty-six-foot cabin cruiser, docked at the Bay Sound Marina on the Chesapeake Bay.

Ron Sr. met with the boat broker, Bill Brown, by himself to get a "crash course" in how to operate the fairly simple, single-engine boat.

"Easy enough. One throttle, one shift," Bill said, tapping on the two levers to the right of the steering wheel. "Turn the key, then push the starter button here."

Ron turned the key.

"No no no!" Bill said. "You gotta clear the bilge of any gas fumes, or it'll be the last time you ever turn that key or any key. This isn't diesel like the trucks you drive; it's gas, and the fumes are explosive."

"Oh, you'd be surprised, Bill," Ron said. "Diesel will blow, maybe worse. Have you ever seen what happens to a jet on TV when it crashes?"

"That's not diesel," Bill said.

"Sure it is; it's kerosine, type 2 diesel."

"Didn't know that. Still..."

"Yeah. The trucking company I work for experimented with it for their fleet of tractor-trailer rigs; it burned cleaner, less wear on the diesel engines, got more horsepower out of them, but crazy expensive."

"Never would've known that. Interesting," Bill said.

"Right," Ron said. "Of course."

"So here. Pull this button out. It turns on the bilge blower, then open the engine hatch and smell for yourself."

Ron followed the directions and, finding the bilge free of gas fumes, started the old Chevy engine; they thew off the lines and took the boat out on the Chesapeake for Ron's first foray as captain of his own boat.

The following weekend, Emily packed a cooler of tuna fish sandwiches, potato chips, and sodas, and she, Ron and Ron Jr. went to the boat for their first family cruise.

When they got to the boat, Ron took Emily and Ronnie by the hand and walked them down the dock to the back of the boat. He didn't say anything, just pointed.

"Oh, Ron! *Really?"* Emily said with a broad smile.

"You bet, sweetheart."

Ron had named the boat *EMILY* and had it painted across the transom.

They hugged, and Ronnie Jr. watched with an ear-to-ear smile on his face.

They boarded the boat. Ron ran the blower, opened the hatch, cleared, and started the engine.

"You sure you know how to drive this thing?" Emily asked as she snapped the child-size orange life jacket on Ron Jr.

"Easy as pie!" He responded. "Ronnie, come here," he said, turning to his son. "I want you to get back on the dock, untie the line on the front of the boat—the bow—toss it on the boat, then come back here, untie that line, and hop on. You got that?"

"Sure do, Dad!" Ronnie responded, full of enthusiasm, his eyes wide.

They cruised out onto the Chesapeake Bay. The winds were calm; the sea smooth, and Ron Sr. felt a lightness in his chest, joy in his heart for the first time in years. He put Ronnie on his lap and let him steer.

"Well, look at that!" Little Ronnie said, pointing to a large yacht coming up the Chesapeake. It passed to the right of their boat.

"Ha!" Ron Sr. said. "That's *EMANCIPATION!*"

"You know that boat, Ron?" Emily asked.

"Of course. That's the presidential yacht! A hundred feet long! It's carried—I don't know—a lot of U. S. Presidents over the years."

"Wow!" Little Ronnie exclaimed. "Maybe someday *we* can own *that* boat!"

"Haha, maybe *you* will, Ronnie!"

The *EMILY* quickly emerged as young Ronnie's ultimate source of joy, an all-consuming passion that forged an unbreakable bond between father and son. With remarkable speed, he mastered the intricacies of boat handling, handling the lines, running the engine, and keeping her meticulously clean.

Whenever Ron Sr. called from out on the road, his greeting was always the same: "So, how's our *EMILY?*" Without fail, Ronnie would retort, a mischievous twinkle in his eye, "You mean our *boat*, right, Dad?"

***

Five months after the purchase, the aged eight-cylinder old Chevy gasoline engine that powered the boat gave out. The ancient cast iron casing had overheated and cracked, dumping gasoline and oil into the bilge, rendering the engine a total loss. Just barely able to afford the boat loan, insurance and dock fees, they had no money to replace the blown engine. And so the boat sat idle. Ron Sr. sank into depression again and upped his drinking.

One day from the road, determined to fight the deep depression that had a hold of him, he called the marina and told them, "Damn it all, go ahead, fix 'er. I want to get her out on the water!" And so the marina mechanic ordered the replacement engine.

***

While Ron was making a delivery of asphalt to San Antonio, Texas, Emily got the call from the marina as she was in the kitchen preparing dinner for Ron Jr. and herself. It was from 20-year-old Rick Anders, a new hire at the marina, having just finished a two-year apprenticeship in marine engines and mechanics.

"Hi, this is Rick at Bay Sound Marina. Is Mr. Bryce there?"

"Uh, no. He's on the road, but I expect him home in two days."

"Oh, that's OK. I'm just calling to tell you we got the old engine out of your boat, the bilge is all cleaned up, and the replacement engine's on its way."

"I'm sorry, what did you say?" She asked.

"We got the blown engine out of your boat, and the new one is on its way. Should be here tomorrow."

"I see," she said, plopping herself down on one of the bar stools at the counter in the kitchen.

"Oh. And the real reason I called is that the marina needs a payment of four-thousand dollars to cover the labor so far, disposal of the old engine, and a deposit on the new installation," he said nonchalantly.

She was speechless.

"Hello?" Rick prompted.

"Oh," Emily said, finally finding her words, "there must be some mistake. We can't pay for that. We don't have that kind of money."

"Look, lady, maybe you better talk to Mr. Bryce. He ordered the engine, and it'll be here tomorrow. So better find a way to pay for it, or I'll just ship it back, and you'll still owe for the labor it took to pull out the blown engine and the shipping and restocking charge for the new one."

"I'll…I'll have Mr. Bryce call you," she said, slowly hanging up the phone.

***

Ron got home from his haul two days later, around dinner time. Emily had already made Ronnie's favorite tuna casserole, and the two of them were at the table when Ron walked in.

"Mmm!" Ron exclaimed, "Your mom's special tuna casserole! I'll sure have some; I'm starved!" He said as he sat at the table and gave Emily a kiss on the cheek.

She recoiled.

Ron knew better than to say anything about it with his son at the table, so he ignored it. The three sat and ate in silence. Little

Ron had noticed his mother's reaction and knew something was wrong.

After dinner, Ron went upstairs to change, Emily and Ron Jr. cleaned up the kitchen, and little Ron went into the living room to watch TV. Emily stayed in the kitchen to put the leftovers away.

Ron came down the stairs straight to Emily and said, "Honey, do we need to talk?"

She told him about the call from Rick Anders at the Bay Sound Marina.

"Is this true, Ron?" She asked as she wiped the kitchen counter, her back to him. "Did you do that?"

"I...I...I thought it was a good idea, you know, for us, for Ron, Jr."

She gave a half-chuckle.

"You know we can't afford that, Ron. There's just no way."

He was well aware of that fact.

"Well, what should we do?" He asked.

"Isn't there only one thing we can do?"

"I don't know, Emily. We can find a—"

"Sell the boat, Ron. We can't do this," She said.

"Sell the boat?"

"Of course," she said.

"No no no," Ron protested. "We can find a way. Maybe your parents—"

At this, Emily lost her patience.

"You keep my parents out of this! This is *your problem,* Ron!"

"But..."

"But nothing! Sell the damn boat! It's either that goddamn boat or me! It's your choice!"

"Oh, come on, Emily," he said as he tried to put his arms around her. "We can—"

"You heard me," she said, "it's me or that boat. You choose."

What neither of them knew was that Ron Jr., having left the living room to get a drink from the kitchen, was standing just behind them and had heard the conversation. He turned back to the living room in disgust.

"Emily! Don't be silly! We'll sell it. No big deal!" Ron Sr. said, putting his arms around her and giving her a kiss on the cheek.

She turned and hugged him.

"Oh, Ron. I'm sorry," she whispered in his ear. "I'm so sorry. You know I didn't mean that."

"I know," he said, "but you're right."

"I love you, Ron. We'll get another boat as soon as we can. I promise."

Ronnie had heard none of this.

Two days later, Ron Sr. stepped out into the dimly lit garage, and in a moment of profound despair, he ended his life with a single bullet.

Ronnie put the full blame on his mother. He would never forget it, and he would never forget their family outings and the sight of *EMANCIPATION. I'll own that boat someday. I'll own that boat! I guess Dad made his choice,* Ron thought to himself. *He made his choice.*

The day he turned eighteen, Ron Bryce, Jr. packed a bag and left his mother, never looking back.

From the house, he hitch-hiked to the Bay Sound Marina with the hope of seeing the *EMILY* one more time. He did. She was concealed in the woods and long abandoned at the far reaches of the marina, set up on cinderblocks, partially covered with a blue rotting tarp, overgrown by weeds; her wooden hull so rotted that half the boat had collapsed into the ground. He climbed onto what was left of the back deck, now on a forty-five-degree slant, and put his hands on the steering wheel, gripping it tightly. In that moment, a surge of dreadful, painful memories flooded his mind, memories of his father's death, seeing him on the garage floor in pools of his own blood, igniting an unbridled rage coursing through every fiber of his being. The *EMILY* now dead before his eyes as well. He squeezed the steering wheel with all his might, his knuckles turning white. "You made the wrong choice, Dad, the wrong fucking choice."

# CHAPTER ONE

**Three Years Ago**
**From Present Day**
**Washington, DC**

Nicolai Volkov was on a mission. Sixty-seven years old, five-foot-six, with a shock of white hair, and built like a former boxer, he sat at the bar with Johnny Stevenson at the crowded Dock Street Bar near the Naval Academy. Stevenson was drenched in sweat, his damp blue T-shirt plastered to his back. He was a thin six-foot-six, having lost weight recently from the stress of what was consuming his life. The air conditioning was failing badly to keep up with the heat and humidity of the day. Condensation dripped off the two bottles of Corona beer on the worn brown Formica bar before them.

"I hope you understand what I'm concerned about," Stevenson said as he picked up his beer and took a swig.

"Of course, I do, Johnny; of course I do, and I'm glad you came to me with this first. You know I take these allegations seriously," Volkov said, "we all do."

"I know you do, Nicolai, that's why I came to you first, you know, before this goes to court."

"I understand."

"She's out of control, Nicolai."

"She is?" Volkov asked.

"Well, she says she's vowed to kill me, kill me in my sleep."

Volkov chuckled. "I wouldn't take that seriously," he said.

"Well, I do. You should hear how she shrieks at me. This has all gotten way out of control."

"That's not good, Johnny, not good. But what's worse is the restraining order she's trying to get against you and the things she's put in that petition."

"You're still living in the house with her, aren't you?"

"Of course. It's my house."

"Not good, Johnny; this is all no good. And her parents are coming from Russia because of this."

"I know, I know, but I never meant to hurt her. Things get out of control so fast, and I just snapped."

"I've seen her petition, John, and the affidavits by her doctor and the social worker."

"I know. It's bad."

"Well, we think we may have found a way to keep it out of court, get you out of this."

Stevenson took a long drink of his beer.

"Thank you, Nicolai."

"We never let our clients down," Volkov assured him. "Never."

Just then, a young couple, who clearly already had a few drinks, came up to the bar and stood next to Volkov.

"Hey, bartender!" The tall, young, athletic-looking blonde girl said, tapping on the bar. "Over here!" She turned to Volkov. "Man, I can't get his attention."

"Well, he's busy, and he's the only bartender here," Volkov said.

"You're sweating," she said.

"Thank you for noticing," Volkov said, smiling. "It's hot out, and it's not much cooler in here."

"And muggy," she said.

Her young male companion, slightly taller than her and broad-shouldered came up behind her, wrapped his arms around her waist and called out to the overworked bartender. He got no reaction.

"I don't think I've ever seen it this hot and muggy since I started at the Academy," He said.

Finally, the bartender made his way over to the group.

"What are you drinking?" Volkov asked the young couple.

"Beer!" She said.

"Four Coronas," Volkov told the bartender.

"Gee, thanks, Mister. I'm MaryAnn Calais. This is my boyfriend, Naval Cadet Randy Emerson," she said with pride in her voice.

"Nice to meet you both," Volkov said.

Stevenson had earlier declined to take a boat ride on Volkov's boat on the Potomac River, considering himself in a precarious situation with Volkov and the whole operation. He didn't feel safe alone with Volkov. That's why he'd insisted on a meeting in a public place.

"Hey," Volkov said, "You guys interested in a nice cool boat ride and maybe a dip in the river to cool off?"

"Oh my God, yes!" MaryAnn said. "You have a boat?"

"I do," Volkov confirmed.

"That'd be great," Randy said.

"Alright then, how about in an hour?" Volkov offered.

"We have our stuff right out in the car," Randy said. "We were going to go to the park. We could go now!"

MaryAnn nodded.

Volkov turned to Stevenson, "OK with you?"

Feeling safe that two others would be on the boat, Stevenson let his guard down.

"Sure."

***

It's against the law to swim in certain areas of the Potomac River. The subsurface currents are so powerful that a swimmer who dares venture into those waters will be quickly dragged down to the bottom of the river, never to be found.

Unless whipped into a fury by the occasional thunder storm, the river's appearance is deceptive with a typically smooth, calm or gently rippling surface—a tranquil facade of placid waters, while invisible underwater currents and relentless tornadic whirlpools of ferocious power silently swirl downward to submerged rock ledges, and life-sucking silt.

Perfect.

He knew the area of the river with the strongest currents and the deepest silt bottoms, and that's where they would go.

The heat and humidity only intensified as they left the bar, piled into Volkov's black SUV, and arrived at the marina and boarded the boat. It was already loaded with beer, vodka, whiskey and an ample supply of pot. Of the four on board the 1951 Chris-Craft Riviera classic wooden thirty-foot mahogany speed boat, two of them believed it was going to be a fun, relaxing, and stoned outing on the Potomac River.

They departed the marina at the Old Town Alexandria Waterfront and took a course south. They passed under the Interstate 495 bridge and onto the wide part of the river. At the helm was Nikolai Volkov. Sitting to his left was Johnny Stevenson; behind them in the back were MaryAnn Calais and her boyfriend,

Naval Cadet Randy Emerson. Stevenson was the target. The others would be collateral damage.

After a leisurely two-hour booze-cruise, Volkov idled the straight-six cylinder engine, put it in neutral, and as the boat coasted to a stop, turned the key and shut it down. With no breeze, the sun and humidity, they were all sweating in minutes. After a few more beers and shots all around, Volkov suggested they all take a swim. "The water's gonna be great!" He said.

MaryAnn and Randy stripped down to their bathing suits and jumped in the water without hesitation; MaryAnn giggling, the two splashing each other happily. After some coaxing, Volkov convinced Stevenson to take off his shirt and shoes, and jump in with his shorts on. The water was refreshing, and he began to relax, doing a slow backstroke.

In a shocking twist that left them utterly flabbergasted, the boat's engine roared to life. Volkov wasted no time; he engaged the gears and slammed the throttle, propelling the vessel forward at full speed. As he stole a glance behind, MaryAnn had already vanished, replaced by the struggling figures of Emerson and Stevenson, grappling with the relentless tug of the descending currents.

With a smug smile, he turned forward again and saw it, but all too late—an ancient railroad tie cast downstream by the thunderous storms of the previous week. At 30 knots, the boat collided with the obstacle, momentarily defying gravity as the massive wooden keel shielded the hull from utter destruction. However, the propeller shaft, making contact with the tie, buckled upward, thrusting the brass propeller, spinning at 3000 RPMs, through the hull like a sawmill blade. The boat ground to an abrupt halt in the water. Volkow was hurled violently forward, propelled through the windshield, over the bow, and into the water. By sheer luck, he managed to snatch a floating cushion that had been ejected into the chaos. The stern of the boat began to fill with water, the weight of the cast iron engine driving its rear deeper into the abyss while the bow pointed skyward. Within a mere minute of the catastrophic collision, the vessel plummeted toward the depths, stern-first, leaving Volkov afloat, clinging desperately to the cushion.

Three hours later, a tour boat carrying a group of Georgetown University students spotted Volkov, diverted from its heading, stopped and picked him up. Volkov was cold and exhausted but had had plenty of time to rehearse his story. The captain and crew of two young women ushered Volkov to the captain's quarters, covered him with a blanket and brought him a cup of hot coffee. Once he was settled, the captain returned to the helm of the boat, dialed up channel sixteen on the VHF radio and called the Coast Guard. They told the captain to hold his position, and they would arrive to pick up the survivor in thirty minutes, as the tour boat was only a few miles from the Coast Guard station.

The US Coast Guard bright orange 25-foot Defender-class boat on patrol in the Potomac River left Hains Point in Washington, DC, flying at forty-five knots—nearly fifty miles per hour toward the tour boat's GPS position. The Defender carries a bow-mounted M240 7.62 machine gun that has an aluminum hull surrounded by a thick, rigid foam collar that makes the boat unsinkable.

As soon as Volkov was transferred to the Coast Guard boat, Ensign Robert Broderick began his questioning, first asking for his identification. Luckily, Volkov hadn't changed into a bathing suit, had never planned to change into a bathing suit, and didn't have one on under his tan shorts and T-shirt. He did, however, still have his wallet in his buttoned back pocket, along with a Russian Embassy badge. He told Ensign Broderick about how the boat sank in perfect detail, knowing full well that if he faked any part of it, someone like Broderick, with his training, would immediately see through it. The only addition to the story was how his three passengers had been thrown from the boat and drowned despite his own "heroic efforts" to save them. He identified his passengers: Johnny Stevenson, MaryAnn Calais, and her boyfriend, Naval Cadet Randy Emerson. He said he'd met all three of them at the Dock Street Bar near the Naval Academy, and when he told them about his boat, they begged him to take them out on the river. Other than that, he knew nothing about them except they were "really nice people."

The investigation would be complicated. Since the sinking took place in federal navigational waters, it would bring in the United States Attorney, the FBI and Homeland Security. And as Randy Emerson was a Naval Academy cadet, the NCIS, the Naval Criminal Investigative Service, would be involved. The NCIS's job is to investigate and crush criminal, terrorist and foreign intelligence threats to the United States Navy—and that includes the Naval Academy.

FBI Special Agent Mary Beth Allison would be the lead agent working with Sandy Mitchell of the U. S. Attorney's office, Assistant United States Attorney, Homeland Security and NCIS. All of this would come under the jurisdiction of the United States District Court for the District of Columbia, Washington, DC.

Volkov met with Mary Beth and Sandy Mitchell voluntarily, for the first time, at the FBI office in DC in the FBI quarters conference room. Mary Beth sat at the head of the table with a thin white folder in front of her; Sandy Mitchell sat to her right, and Volkov sat across the table from her. Being in charge of the investigation, Mary Beth led the meeting and did the majority of the questioning. Volkov had no fear of her or Sandy Mitchell. They were both soft-spoken respectful, and appeared to empathize with the "tragedy he had suffered." He found them both highly attractive, both dressed professionally—Mary Beth in a navy blue skirt and jacket and impeccable button-down white shirt, two-inch black heels; Sandy in a black version of the same suit, gray shirt. He hadn't noticed Mary Beth's Glock 9 semi-automatic pistol in the holster under her jacket. Nor was he aware of the massive power the two possessed in the law.

He rattled off the same story about the sinking. When Mary Beth asked to see his Russian Embassy badge, he said he'd left it at home. Sandy Mitchell raised an eyebrow at that. Mary Beth had already gotten the intelligence on MaryAnn—Senior at Georgetown University, hometown Philadelphia, PA, well-to-do upper-class family, dad a family medicine doctor, mother a third-grade school teacher, no siblings. She was apparently just out for a good time the night before the sinking with her boyfriend, Randy Emerson, the Naval Cadet. They encountered Volkov and Johnny Stevenson at the Dock Street bar, and they all got talking about the summer heat and humidity and, eventually, Volkov's boat. Volkov told them he'd met Stevenson for the first time at the bar that night and knew nothing about him.

The meeting ended after only an hour since Volkow apparently knew nothing of the three that perished in the accident. Neither Mary Beth nor Sandy Mitchell were convinced. There being no reason to hold Volkov, they excused him.

Despite a diligent search, Mary Beth drew a blank on Johnny Stevenson, so she sent a U. S. Marshall to pick Volkov up for another interview. He was nowhere to be found, and the Embassy denied any knowledge of his very existence.

Six months later, Volkov's boat was recovered.

Eight months later, in the middle of the investigation, and much to Mary Beth's frustration, she was transferred up to New York State to join in an investigation of a murder-for-hire continuing criminal enterprise.

MaryAnn's, Johnny's and Randy's bodies were never recovered.

# CHAPTER TWO

### Present Time

It was a perfect late October autumn Saturday. The leaves were at their peak colors of crimson, yellow, orange, and deep purple, and the sky was cloudless, clear and blue. Vince was dressed casually in khakis, a black sweater and Dockers deck shoes. His dark brown hair, just recently starting to show streaks of white, was tussled, and he didn't care. His decades as a trial lawyer manifested itself in slight wrinkles in his otherwise smooth, clean-shaven face, and despite the stress that is an everyday part of that practice, he managed to maintain a slim and fit appearance. It was his day to be with the girl he loved, and away from the law office and his usual neat pin-striped suits, white starched shirt and tie.

*Ah, I've finally reached a place where I can just relax and enjoy life,* Vince thought as he drove his Tesla down interstate 81. He turned to Mary Beth in the passenger seat and smiled. *All is good,* he thought, admiring her beauty.

Vince had turned most of the law firm's cases over to his son and partner, Michael, with the hope of being able to take more time to enjoy some leisure after nearly four decades of practicing law. It was his dream to spend more time with Mary Beth and appreciate the better things in life with her.

"You look so beautiful," he said.

Mary Beth sat in the passenger seat of the car with casual confidence, her long, slender legs gracefully crossed at the knee, every move understated elegance. She sported impeccable pristine white sneakers, jeans that clung to her form, a plush navy-blue sweater and a flowing camel coat that exuded sophistication. Her long, cascading brown hair, interwoven with threads of gold and auburn, was parted on the left, perfectly framing her slender face and flowing over her shoulders in waves. As she turned to Vince, the radiant sun pierced her beguiling, hazel eyes, capturing Vince for a moment.

"So, what's the name of this place we're going?" Mary Beth asked.

"Bay Sound Marina, Annapolis," Vince said, liking the sound of it as he did.

"Bay Sound Marina. Hm." She said, thinking it sounded a bit familiar.

"It's a covered marina, so that's where a lot of classic and antique yachts are kept."

She gave him a beautiful, broad smile and leaned back in the seat. "How long until we get there?"

"You mean, are we there yet?" Vince said with a chuckle. "You sound like Mike when he was five years old. Just about two hours."

"Funny guy," Mary Beth said, not at all amused. "We've been on the road for four hours. I'm starved." She thought about the Snickers candy bar in her purse.

"OK. Let's find a place to eat. We're almost to Harrisburg. I know a good place."

"Perfect."

***

Vince and Mary Beth had met each other when Mary Beth was a Special Agent with the FBI investigating a murder-for-hire syndicate undercover. Vince DeMarco, thirty years her senior, was a seasoned trial attorney practicing with his son. Her case led her to investigate individuals that directly and indirectly led to Vince's past clients and their contacts; some involved in the conspiracy; some were targets. She was finally assigned to protect Vince and his son, Michael, from the syndicate itself that had targeted them. During her undercover stint, she and Vince had fallen in love. Despite her disclosure to her supervisor of the relationship, which was strictly against FBI policy, she was allowed to remain and complete this one last case to be sure Vince and Michael would finally be safe. When Vince came under a violent attack, Mary Beth managed to save his life. After that, and after blowing her cover and revealing herself to Vince, she resigned from the FBI. That was a year ago. Since then, Mary Beth had been acting as a consultant to major corporations, private security companies, and defense attorneys, including the DeMarco Law Firm, LLP. Despite being a licensed attorney, she chose not to practice law. She and Vince had become bonded for life. She worked when she felt like it.

***

Twenty minutes later, they took the exit for Harrisburg, and Vince found his way downtown to North Street and Armando's Italian Restaurant. It was a bistro one would more expect to find on Mulberry Street in Little Italy in New York City—long and narrow, tall ceiling, with small round tables, covered in red and white checkered tablecloths in an old red brick three-story building. The aroma coming from the kitchen in the back was sweet and inviting. Vince helped Mary Beth out of her coat and placed it over the third chair at the table. As soon as they sat, a young boy, no older than twelve years old, wearing a white dress shirt, black bow tie, black slacks and a crisp white apron, approached the table with two glasses and a pitcher of water, set the glasses in front of Vince and Mary Beth, poured the water, turned on his heels and left. He returned a minute later, handing them each a paper menu and left.

"Oh my God, he's adorable," Mary Beth said, watching the boy as he disappeared into the back of the restaurant.

"The whole operation is a family business," Vince said. "His grandparents immigrated from Rome, Italy in the 1950s."

"That's wonderful. How do you know them?"

"Ah. Well, remember my dad, Santino, was an immigrant from Rome, too. When he came over, he was already an attorney, and he did a lot of immigration work, especially for families he knew back home. He did the work that got Armando Cellini and his family to the U.S. We've known them ever since. Now, the place is run by his son, Armando, and his family.

Just then, a tall, slim, jet-black-haired man dressed in an outfit that matched the little boy's approached the table. Vince stood, and the two embraced.

"Vince! It's so good to see you!"

"Armando, I'm happy to be here." Vince turned and took Mary Beth by the hand, "This is Mary Beth."

Armando took her hand from Vince's and gave it a kiss. "Ma, Che Bella!" he said.

Mary Beth blushed, "Well, thank you," she said demurely.

"Sit, Vince, sit!" He said.

Vince handed him the two menus. "We don't need to order, Armando. You just bring us what you think best."

"Certo!" Armando said, giving Mary Beth a slight bow. He left for the kitchen.

"That's pretty brave," Mary Beth said.

"Naw. Not at all. He knows exactly what I like."

"Oh really? What about me?"

"You'll see."

Armando returned in two minutes with a bottle of 2019 Masseto Toscana wine, two glasses, opened the bottle and poured.

Mary Beth squinted her eyes, reading the label. "Vince," she said.

"I know, I know."

"Vince, that's a thousand-dollar bottle of wine," she said in a whisper.

Armando heard her and winked at Vince.

Vince changed the subject. "So, do you want to hear more about her?"

"Who?"

"Her. What we're going to see."

"Oh. The boat," she said flatly, then taking a sip of the wine. "Sure."

Vince's eyes lit up. "Well, she's fifty-three feet long, with a sixteen-foot beam, 1969 Matthews...."

"Yes, this much I know, and all wood—fifty tons of wood."

"Wood. Right, and..."

"And it's only forty-thousand dollars." She said, taking a deep breath.

"Yes, uh. Yes."

"Why so cheap? I don't know a lot about boats, but I do have a computer, and the boats like this one are selling for over a quarter of a million dollars."

"Well, that's why we're going to see her. It's a hell of a bargain."

Mary Beth took a long sip of wine. "A bargain, I'm sure."

Just then, Armando appeared with two plates of Bucatini all'Amatriciana.

"This looks wonderful," Mary Beth said.

"Yes, it's..."

Mary Beth cut him off. "spicy sauce, black pepper, dried chiles, guanciale–that's pork—delicious."

Armando listened, nodding his head slightly, raised his eyebrows, gave Vince a side-glance, and left.

"Well, bon appetit," Vince said, digging into the delicious food on his plate, twirling the thick pasta with his fork.

The conversation paused as they both savored the food and wine, occasionally just smiling at each other.

"Remember the art exhibition at the Statler at the Bar meeting?" Vince asked Mary Beth.

"The Bar meeting? You mean the Bar Association dinner we went to with Mike?"

"Yeah. And remember meeting the artist? Rose was her name; I'm pretty sure."

"Of course I do," Mary Beth said. "Yes. Rose."

"Well, apparently, she and Michael hit it off pretty well."

"I did notice they seemed to have quite a long conversation that night."

"Mike tells me they're dating."

"Oh really? They just met two weeks ago."

"Yes, really."

"Michael and an artist." Mary Beth said. "Hm."

"And why not?" Vince asked.

"Well, it'll be good for him to see a part of life that isn't the law; expand his horizons a little. And I think she's adorable! Petite blonde with those big blue eyes, very pretty."

"Something new for this family," Vince said.

Mary Beth gave Vince a quizzical look. "What?"

"Yes. An artist. I mean an artist."

"Oh. Yes. So she is."

"And she's not a lawyer."

"Right. She's not a lawyer. Well, I can't wait to see more of her work," Mary Beth said.

***

The Chesapeake Bay was smooth and calm and perfectly mirroring the blue of the sky. Small wavelets lapped against the old concrete bulkhead of the Bay Sound Marina. It was a covered marina with three piers of eight slips each—24 boats—under a corrugated steel roof held up by only wooden beams at each corner and every fifty feet along each pier. Other than the roof, it was all open-air. It was host to some of the most rare and pampered antique and vintage yachts in the country. Even the fuel the marina provided them was exceptional, and one of the only marinas in the country that provided it: Type 2, diesel, also known as Jet A fuel-the same used in modern airliners. It burned cleaner, extended the life of the old diesel engines and caused the old engines to produce ten percent more horsepower than regular marine diesel would. The only downside: it was highly explosive under the right circumstances.

The two sat across from each other at a small round teak table in the salon of William Brown's fifty-foot Chris-Craft wooden cabin cruiser—two beers on the table. Despite the crisp, fresh air and light breeze outside, the inside of the boat was musty, with the lingering smell of spent cigars and alcohol. Brown was the yacht dealer who handled all boat sales at the marina. A retired IBM executive, he'd been living on his boat year-round for over twenty years. Tall, once distinguished, with thinning gray hair, and wearing slightly tattered blue jeans and a red and blue striped Polo shirt tucked into his pants, Bill believed he was dressed his best for meeting Vince and Mary Beth. Rick Anders was short, stocky with a full head of gray and red hair, dressed in oil and grease-stained jeans and a gray sweatshirt equally stained with specks of dry paint and epoxy. He smelled of diesel. He was the marina mechanic and general "fixer."

"So, they're coming today?" Rick asked.

"Yeah," Bill said. "On their way now."

"What have you told them about the boat?"

"Not much."

"Nothing about the history?"

"Just the good stuff," Bill said.

"OK. Good. This guy know much about boats?"

"He's a lawyer for chrissakes."

"That doesn't mean anything."

"Other than the fact that he said he once owned a thirty-six-foot cruiser, I don't know."

"Well, we just have to get that thing out of here. She tells too much of a story."

"No shit," Bill said. "You worried he's a lawyer?"

"Yeah. We'll see. He seemed so over-excited on the phone; this might be a quick sale."

"Let's hope so."

"What about the girl?" Rick asked.

"I talked to her briefly on the phone when he wasn't there. She wasn't enthused."

"You know anything about her?"

"Nothing. Sounded nice on the phone. Young"

"Everybody's young to you," Rick said.

"You're not." Bill snapped.

There was a long pause as the two took time to drink their beer.

"Did you air out the Matthews?"

"Yeah. I opened all the doors, hatches, and windows; did what I could."

"Ok, Good," Bill said. "Engines?"

"Engines run, generator's dead."

"Can you get it to run at all?"

"Look, I fix 'em; you sell 'em. Generator's dead. By the way, how's the new widower doing?" Rick asked.

"He's good. All settled in. He'll be at the meeting Friday. Said he's never been happier in his life."

"Hmm. Well, I'm outta here. See ya later."

# CHAPTER THREE

### Annapolis

Vince and Mary Beth were a bit surprised to see the man out in the parking lot of the Bay Sound Marina waving to them vigorously.

"I guess that's the guy," Vince said, pointing.

"I would think so," Mary Beth said. "I don't imagine anyone else here that would be so happy to see us."

Vince gave a chuckle. "Well, let's go meet him."

"I can hardly wait," Mary Beth said, hoping the guy wouldn't go in for a hug.

Vince parked, and they got out of the car. Bill approached them quickly. It made Mary Beth uncomfortable.

"Well, how was the drive?" Bill asked.

"Very nice," Vince said. "Beautiful day for it."

"Beautiful indeed," Bill said, staring at Mary Beth. She instinctively turned her head away.

Bill smirked. "Well, let's get right at it, shall we?" He said. "Right this way." He pointed to the pier behind him. "She's all the way at the end."

As they walked down the pier, Mary Beth was taken aback by the smell of diesel fuel in the air and rubbed her nose. Vince was loving it. He noticed her grimace.

"That's diesel, honey," Vince said.

"Yeah, these are mostly all diesel-powered boats," Bill confirmed.

"And a lot of wooden boats," Vince said.

"Yeah, we got mostly older wooden boats here. They're *all* wood. It's a covered marina. Keeps em' better than out in the open."

Vince nodded. Mary Beth didn't care.

"Well, there she is," Bill said, pointing the boat in the last slip on the pier. "I have the engines running for you. They run real smooth."

"It's big." Mary Beth said, waving the diesel exhaust away from her face.

"Fifty-three feet," Bill confirmed.

Mary Beth looked at the transom of the boat that spanned all of sixteen feet across. "*MY CHOICE,*" she read aloud. "Odd name for a boat, isn't it?"

"Naw," Vince said. "You'd be surprised that names people give their boats."

Bill stayed put. Vince took Mary Beth by the hand, and they walked down the dock, eying the boat over.

"It leans to the left," Mary Beth said. "Should it lean to the left?"

"Ah. She's got a critical eye, this one," Bill said.

"You have no idea," Vince said.

"Well, yeah, I see she's listing to the port side," Bill said. "Maybe there's more fuel in the port tank. That'd do it." Bill said with a shrug.

They climbed up the boarding stairs to tour the old boat.

"Once belonged to a congressman up in DC," Bill said.

"Yeah?" Vince said, inspecting the main salon then walking down into the galley.

Mary Beth started to take a seat on the small couch in the salon but then thought better of it. The smell of fuel was strong inside the boat. She stood with her arms crossed in front of her. Bill ushered Vince throughout the boat, then opened the floor hatches to show him the two massive Cummins diesels. The smell of diesel was so strong, Mary Beth excused herself, left the boat and waited on the dock. She was sure her camel coat would forever now smell like an old boat.

There was another smell in the boat, a faint but putrid smell that she couldn't identify. She was glad to be out of there.

They all walked back down the pier to Bill's Chris-Craft.

"Step into my office," Bill said, pointing to the boarding stairs to his boat.

Bill had set up a small folding table and three chairs on the open back deck of the boat. Mary Beth was relieved. She was in no mood to step down into another old wooden boat. He motioned to the chairs.

"Either of you care for a drink?" Bill asked.

Mary Beth just shook her head.

"I'm good," Vince said. "Thank you."

"Alright then," Bill said, "Let's get down to it. How'd you like the boat?"

Mary Beth had her opinion, but kept it to herself…for now.

"She needs a lot of work," Vince said, with his arms crossed in front of him.

"Of course she does," Bill confirmed. "Why else would the price be so ridiculously low?"

Mary Beth tightened her lips and looked over at Vince. To her dismay, he had a big grin on his face. *Oh my God*, she thought, *he looks like a little kid about to get a new bicycle*.

Bill chimed in. "You get it up to your marina in Ithaca, and I'm sure you can get her fixed up just great. You know, a Matthews like this all polished up pretty goes for well over a couple hundred thousand."

Vince nodded his head, with raised eyebrows and a grin. Mary Beth figured it was pretty much a done deal, but she still had hopes Vince might break out of his stupor.

Bill picked up a grimy manilla folder on the table and handed it to Vince. "Here's the documentation and some records of repairs that have been done over the years."

Vince did a cursory glance through the papers. The top sheet was a bill of sale dated that day in his name with a price of $40,000.00 as if it was already a done deal. The second was the dock rental agreement in the prior owner's name. Behind that were three faded color photos of the boat at sea. He pulled them out and showed them to Mary Beth.

"Isn't she beautiful!" A statement rather than a question.

"She *was*." Mary Beth said flatly.

He scanned the rest of the file and noticed the last repair appeared to have been done just two years ago.

***

**Ithaca, NY**

The isolated one-story, white, aluminum-sided 1960s ranch house surrounded by tall pine trees sat on the side of West Hill of Ithaca, New York, overlooking Cayuga Lake. Dr. Raymond Weber, M.D. had lived there alone since his divorce over twenty years ago.

It had been a brutal divorce. He met his now ex-wife, Helen, when he was in his last year at the State University at Binghamton, finishing up his BA degree in pre-med, and they married that summer after his graduation. Helen had already completed her RN degree and had taken up a job at the Binghamton Clinic. She would provide the only income for the couple until Weber finished up med school and obtained his license to practice. After twelve years, she could no longer tolerate his drinking and abuse and filed for divorce. The court was quick to award her hefty alimony, half his retirement, as well as an equitable distribution of half the value of the medical degree he'd earned while she supported him.

When he started medical school, he had dreams of a brilliant, successful psychiatric practice. His first two years went well, and he had a good shot at realizing his dream. He'd leased a two-room office in a new office building in downtown Ithaca and had a receptionist in the front office behind an oak-curved desk in charge of scheduling his patients' appointments and making sure they paid before they left the office. His office was plush with a two-thousand-dollar tan Herman Miller upholstered chair and matching Boudreau over-stuffed leather sofa for his patients, a polished oak coffee table, plush tan cut-pile carpet, and a wall of oak bookshelves filled with a selection of books on mental health, therapy techniques, and biographies of famous psychiatrists. By the third year, however, his patient list began to shrink, as although he had plenty of new patients making appointments, they, more often than not, after spending a session or two with him, never returned. He could no longer afford to pay his receptionist, and after missing two paychecks and Weber blaming her for the failure, she quit. He fell behind in his rent. The stress led him to drink. And drink. And drink more until he was drunk almost every night. With that came missed appointments, poor reviews, more alcohol, complaints to the education department that he appeared intoxicated in sessions, and finally, the third malpractice lawsuit against him. His complete hate of life gave him a hair-trigger-raging temper that all but assured his failure. He lost his practice as he became well known for his alcoholism and bad temperament, so he took a job with the county social services office as the in-house shrink.

Instead of backing him up, the county he worked for told him he was on his own and there'd be no support for him in the pending malpractice action, financially or any other way. He was lucky to keep his miserable job.

Lately, the loneliness had become more than he could tolerate. Being sixty years old, 5'4" tall, balding, and well over 250 pounds with no social skills, he'd had no luck dating. So, two years ago, he began his web search for a new wife. He had only one main criterion: that she be desperate enough to marry the likes of him.

He'd given up on Match.com, e-harmony, Ourime, and the other standard dating sites that have gotten only mocking and insulting responses to his posts. After searching the internet for hours one night, he found "Russianbryde.com." Unlike any of the other dating sites, Russianbryde *guaranteed* to produce a "young, beautiful, vivacious bride." Also, unlike the other dating sites, this one required an up-front "retainer" of $25,000, refundable if a bride was not produced within sixty days. He reviewed the hundreds of photos and postings of girls until the early morning hours and landed on Alina. Her photos were true to herself—modestly dressed; some photos of her in a floral dress, barefoot on a swing in the yard, some with her oversized glasses on, most without. The only slightly provocative photo was of her in a modest white two-piece bathing suit, sitting on the grass by the edge of a pond with a flower in her hand, her long legs crossed at the knees. A photo of her leaning against a shade tree showed her sandy blonde hair flowing long, well past her hips, a tall, thin girl with big, trusting hazel eyes.

Alina lived in Tolyatti, a city known to have the highest poverty index in Russia and had heard about the site from other girls in her town. She looked it up and saw it guaranteed the "men seeking women" had been thoroughly vetted for finances, criminal records, prior marriages, and were even subjected to psychological testing to assure they were "safe". The site included a "guarantee" that if there was any danger to the women that used the site, they would cover the costs of relocation back to their home in Russia "from anywhere in the world." The site included an "emergency phone number" in case the girls ran into any trouble.

Weber responded to Alina's post. He was completely taken by her looks; she was completely taken by the letters "M.D." after his name. After weeks of emailing, then texting, then FaceTiming, Weber charged the twenty-five grand to his Amex and agreed to wire her $1,000.00, send her a one-way ticket to New York, and upon her arrival, to send her family $10,000.00.

She arrived promptly. He found her to be shy, delicate, beautiful, and sweet-natured. She found him to be disgusting at every level.

A month after her arrival, they were married by the local town justice.

She quickly took on her "wifely duties." She cooked, cleaned, and tended to his "needs." She had driven her mother to do errands and doctor's appointments back in Tolyatti, so she knew how to drive. When she asked Weber if he would teach her how to drive in the U.S., he flatly refused. As far as he was concerned, she had no reason to leave the house—ever.

Months passed.

She had set the table for dinner and had made his favorite dinner: sauerbraten and sauerkraut. They sat down to eat.

"Can I ask when you will send the money to my family?" She asked in a soft tone as she served him.

He was already good and drunk. No response.

She paused and watched as he devoured a huge piece of the sauerbraten, chewing open-mouthed.

"So, can I ask?" she started.

Weber abruptly stood, threw his plate across the room, and screamed, "SHUT THE FUCK UP! YOU'RE HERE, AREN'T YOU? AND I MARRIED YOUR SORRY ASS, DIDN'T I? NOBODY'S GETTING ANOTHER DIME FROM ME!"

He stormed over toward the front door, grabbed his coat off the rack, and started to open the door.

Alina ran to him, "I'm sorry! I'm sorry! Please no! Please don't go! I'm sorry!"

And with that, Weber back-handed her; she fell to the floor sobbing, holding her face, and he left.

# CHAPTER FOUR

## Annapolis

Bill Brown sensed a bit of tension between Vince and Mary Beth. Clearly, Vince was in the bag and wanted that boat; Mary Beth's vibes were just the opposite.

"If you two need time to talk, I understand. There's a nice diner just down the road from here, about a mile."

"Yes. We saw it on our way in," Vince said. He turned to Mary Beth. "You wanna go?"

"We should talk," she said softly, smiling at Vince.

"OK then, let's go," Vince said. He turned to Bill. "We'll be right back."

That surprised Mary Beth.

"Oh, here, take this," Bill said, handing Vince the manilla folder.

***

Vince and Mary Beth arrived at the "Sea Star Diner" ten minutes later and sat at a booth by the front windows. They both ordered coffee.

Mary Beth held her cup with both hands, trying to warm herself. She waited for Vince to start the conversation. He thought hard.

"Well, for one thing, it could be a good investment for us," he said.

That stunned her. "Investment? Really, Vince? Since when has a boat ever been a good investment?"

This was the Mary Beth Vince loved so much: strong, intelligent, and almost always right about everything.

"Well, you know…" Vince started.

"Vince, why don't you just tell me what you're thinking, how you feel about this? The money isn't the issue; it's just that it's a lot of boat, a lot of *wooden* boat, and it needs a lot of work."

Vince nodded his head slightly, took a sip of his coffee and thought about how to respond.

"From the heart, Vince; don't go all lawyer on me." Mary Beth said, then gave him a small, closed-lip smile. She leaned back in the booth and sipped her coffee, still holding the cup with both hands.

"I need a challenge, Mary Beth," Vince said. "Since I handed most of my caseload over to Mike, I've felt a little lost."

"I understand. That's what I thought," Mary Beth said. "I understand, Vince. You can't go from a hard-hitting trial lawyer for forty years to just sitting around. I get it."

Vince tried to not get emotional, but he was losing the battle.

"And as much as I love you, Vince, it's been getting harder on me, too, since I left my job at the FBI. I mean, our leisure time together is wonderful; it's a dream, but we're just not used to it. I don't think we're cut out for it."

"Oh, how you get me, Mary Beth," Vince said with a subtle smile.

"You know I do, Vince." She said. "So, this is how you see this old boat? A challenge?"

"I do."

"Then I'm all for it, Vince, and I love you so much."

***

**Ithaca**

Susan and John Cranch had three children: Marie, age five; Alfred, age seven; and John Jr., age ten. Susan had accused her estranged husband, John, of inappropriate physical acts with Marie and that he'd been physically abusive to the boys. John held steadfast in his denial and accused Susan of parental alienation of the children against him. After six court hearings, a dozen witnesses, four temporary custody orders, and two failed efforts at mediation, the family court judge decided he would make his final custody order based on the testimony of one last witness. So, Susan and John Cranch were ordered by the family court judge to submit to psychiatric evaluations in their hotly contested custody battle.

Weber's office was on the second floor of the county social services building. The room was small, just large enough for his worn oak desk and stained green fabric office chair, and two mismatched metal and fabric chairs for his patients. There was no plush leather couch, vast ornate mahogany desk, and Herman Miller office chair that once made up his private practice office.

This was a joint session for the parents and would be the last one. They'd each submitted to individual evaluations by Weber and had now come for the last meeting with Dr. Weber to discuss his report to the judge. Susan had taken a dislike to Weber from the first time they met, and it had gone downhill from there; John, on the other hand, felt some camaraderie with him.

"Have a seat," Weber said as the two entered his office.

Susan grimaced as the stench of alcohol emanating from Weber hit her; he was drunk.

As Weber fumbled with the manilla folder on his desk, four sheets of paper slid out and fell onto the floor. He left them there.

Susan sat up straight on the edge of her chair; John sat back, his legs crossed at the knee.

It was clear Weber was struggling to focus.

"So, where are we?" John asked.

Weber cleared his throat twice, "We're about done," Weber said. "I've already made my recommendation to the judge."

John had a smirk on his face; Susan's mouth went dry.

"Dr. Weber," Susan started," I'd like to address some of the things you said to me in my last session with you."

Weber chuckled. "I bet you would, little lady," he said.

"Well, you accused me of coaching the children against their father and…"

"Oh, don't even bother," Weber said, cutting her off. "You're a disaster for these kids. You've been coaching them all along."

Susan gasped. "I don't understand where this is coming from!" She said, holding back tears. "I've done no such thing."

"You've completely alienated them from John. They don't even want to see him at all," Weber said.

"Are you trying to say that I forced them to make those horrible things up and that I somehow convinced the school to report to Child Protective Services that he abused them? That I told them to make it up?" Her eyes began to water, and she felt her throat constrict.

"That's exactly what I'm saying," Weber said, slightly slurring his words. "You women do that all the time, just to get custody of your kids so you can get child support. I know the game you're playing."

"Oh my God!" Susan exclaimed, "Who are you? What kind of person are you?"

"I'll tell you what kind of person I am," Weber said, "I'm the kind of person that's told the judge that you're a danger to your children and that you shouldn't have custody or even visitation with them."

"No! No no no!" Susan said, now sobbing. "No, please!"

"And starting today, John will pick them up from school, and he'll have sole custody, and I guess you'll have to go back to court if you want to see the kids again. And good luck with that."

"Oh my God! How did this happen?" Susan cried.

"The judge left it up to me, remember? I filed my report yesterday; he issued the order this morning."

Susan was in complete shock.

"Please, please, please, don't do this to my children, please!" She said, gasping for air as she wept.

"Oh boo-hoo," Weber said, winking at John. "We're done here. Get out."

***

After that fateful session, she returned to the solitude of her bedroom. In her trembling hand, she held a prescription bottle of 30 Xanax tablets, a remedy prescribed for her by Dr. Weber to calm her down. With a heavy heart and a bottle of wine by her side, she swallowed those pills one by one, chasing each one down with the wine. She lay down on her bed, and within the hour, she fell into a dark, unending sleep.

***

Later that year, John Cranch sued Dr. Weber for malpractice on behalf of his children, claiming he was the cause of his wife and the mother of his beloved children, Susan's death.

Weber's malpractice insurance company hired a large medical malpractice defense law firm in New York City to defend the lawsuit. Sean Adams, six-foot-three inches, handsome, jet-black-haired, blue-eyed thirty-something, a Cornell Law grad working "of counsel" for that firm, would handle the preliminary matters of the case, including discovery and depositions.

Adams met Weber's wife, Alina, a few times while representing Weber. He was smitten with her almost immediately. To him, she was the most beautiful, delicate, kind girl he'd ever met. But each time, she'd shown evidence of abuse. Not just the swollen lip and black eye Weber explained as her being a "klutz" and falling down, but just as revealing was the way she'd diverted her eyes from him and remained silent in Weber's presence. He'd grown to hate his client more than he ever thought he could hate any human being.

Their first encounter was over dinner for Adams to go over the court-ordered trial calendar with his client, Dr. Weber, and to discuss the upcoming depositions. He didn't intend to go into anything detailed with his client that night and would schedule another meeting to prep Weber for his depositions.

Adams was already seated when Weber and Alina entered the restaurant dining room. He stood as they approached and held the chair for Alina. She gave him a demur smile, tucked her skirt under herself, and sat.

"This is my wife, Alina," Weber said.

Adams was surprised Weber brought his wife to the dinner meeting. He reached across the table and took her hand in his. "Nice to meet you, Mrs. Weber," he said, noticing how warm and delicate her hand was, her grip almost nonexistent.

The server approached the table and asked if anyone wanted anything to drink besides the water on the table. Weber jumped right in.

"I'll have a vodka martini," Weber said, "dirty, olive juice, a drop of vermouth, and make it a double."

"And for the lady?" The served asked, turning to Alina.

"She'll just have water," Dr. Weber answered.

Alina's eyes met Adams' as if to be sure he saw what had just happened. Adams' eyes widened.

It was at that moment Adams noticed the slight swelling and fading bruise over her left eye. She reached up to her forehead and touched it with two fingers, her gaze holding his. It put a shiver down his spine.

Adams ordered a bourbon on the rocks.

"You're paying for dinner, I assume," Weber said. "I mean you or the insurance company, right?"

Adams shook himself out of the haze, "Yeah. Sure. Yeah."

"OK, good," Weber said, nodding, and signaling to the servicer. "Look, I'll have a shrimp cocktail, the filet mignon—rare—the Maine lobster, garlic mashed potatoes, green beans, and get us some bread and butter."

"Of course, sir," the server said, "and for the lady?" he asked, turning again to Alina with a bit of trepidation.

"She'll have a small Cobb salad," Weber answered in her stead.

Adams ordered salmon and a salad.

Weber ordered another double vodka martini.

Throughout dinner, Alina's and Adams's eyes met, sometimes lingering.

After Weber's third double vodka martini, he became belligerent with both Adams and Alina, calling Adams incompetent, and when Alina was about to object, raised his hand as if he was about to strike her. Adams called for the check, paid it and left. He feared for Alina.

The next time they met was at Weber's house on a Saturday afternoon, this time to prepare him for his upcoming deposition. Adams and Weber sat at the dinette table, documents scattered across the top. Alina was nowhere to be seen. It was one o'clock, and Weber already reeked of alcohol. Adams reviewed the complaint against Weber, their answer and affirmative defenses, demands to produce, and expert reports—all looking very bad for Weber. As Adams droned on through the documents, Weber's temper mounted.

"This is bullshit!" Weber said.

"Well, actually, Dr. Weber, we need to—"

"Alina!" Weber called out. "Alina! Get out here!"

The bedroom door opened, and she scampered out into the hallway, "What do you want?" she asked.

"My drink! Make me my drink!" he demanded.

She knew exactly what he meant. "His drink" was two shots of vodka and a half-shot of vermouth.

She looked at Adams. "Now?" She said to Weber.

"Fucking NOW!"

A minute later, she emerged from the kitchen with a glass in her hand and handed it to Weber.

Adams looked at her in shock. Her lip was split in the right corner and still bleeding; her right cheek was swollen and dark.

"OK," Adams said, standing and gathering his papers. "That's all for today."

His eyes locked with Alina's.

"Fine," Weber said, taking his drink and walking, stumbling, into the kitchen.

"I'll help you with your papers," Alina said to Adams.

"No, that's OK, I got it," Adams said.

She picked up the pen on the table and wrote, "HELP ME," on his yellow legal pad. He nodded. He left.

The third and last time Adams and Alina met was days later. Adams first checked that Dr. Weber was at his office by calling him with some unnecessary questions about his malpractice case, then showing up at the Weber's house unannounced. He knocked on the door, and Alina answered.

"He's not home," she said before he could say anything. She kept the door opened only a few inches.

"I know, Alina. I'm here to see you," he said, applying slight pressure against the door.

"No, please, no. Just go away."

He put more pressure against the door until she relented and opened it. He reached out to hug her, and she pushed him off with a stiff arm to his chest. "No. Don't." She said firmly.

"Alina, I just want to tell you, I'll do anything for you, I'll do anything to help you!"

She shook her head, turned her back to him, and took two steps into the house. He followed her closely, took her hand, and turned her around facing him. She pulled her hand away.

"You have to leave. Please leave now," she insisted.

"Alina, I love you! I would do anything for you, anything!"

At that, fear and panic rose in her, and she pushed him backward toward the door with both her hands on his chest. "Go! Go now!" She said. "You must be crazy!"

She continued to push him with all her weight and strength until he was at the threshold of the still-open door, then gave him one thrust backward. "GO!" She yelled, "GO!"

He turned and walked out the door. "You'll see," he said. "You'll see."

***

Weber's deposition did not go well for the doctor in the case of *John Cranch o/b/o the Minor Children, M. Cranch, A. Cranch and J. Cranch, Plaintiffs v. Raymond Weber, MD, Defendant.* How could it? Weber was being sued with allegations that while treating a thirty-year-old woman and her soon-to-be ex-husband going through a divorce for "co-parenting problems," he had regularly demeaned her, insulted her, made derogatory sexual comments, blamed her for the failure of the marriage and opined she was a danger to her children. The lawsuit also accused Weber of being intoxicated during sessions—a claim that had been raised in several complaints against him before—and that he had over-prescribed her Xanax. Unfortunately for the doctor, he had dutifully recorded these sessions that would be fully admissible at trial. Worse for him, Susan had kept her friends apprised of all this through emails and FaceBook posts, and unbeknownst to Weber, she had recorded a few sessions herself where he was visibly intoxicated and abusive—also admissible at trial.

Now, John Cranch was suing Dr. Weber for malpractice on behalf of the children and for his own "extreme emotional distress" arising out of Susan's suicide.

Dr. Weber's deposition was held on the second floor of the courthouse in the room usually occupied by the grand jury. He sat next to his lawyer, Sean Adams, and across from the plaintiffs' attorney Jack Stevens who questioned him relentlessly about his record, and how he treated Susan. Weber was uncooperative, vituperative, sarcastic, and finally after six hours, stood and launched into a tirade of expletives so loud the court bailiff came knocking on the door to quiet him down as people in the courtroom could hear him. The deposition ended, and Weber left the conference room with his attorney, Sean Adams.

"Let's go back to my office to talk, Doc," Adams said.

Weber sneered. "Go fuck yourself."

Adams shook his head. "Look, this went badly, and we need to talk."

"Why bother? You just sat there like a fucking moron, no objections, nothing. You threw me under the bus. You're worthless."

"It's a deposition, not a trial. I can only object to the form of the question, and her questions were not objectionable."

"I should've gotten a *real* lawyer," Weber said. "I don't know what the fuck you are."

"Listen," Adams said, "You've got to cooperate with me! You're being sued for ten-million-dollars, and you've only got one-million in malpractice insurance coverage. I've reviewed your disclosure of assets; you don't have much. Just your house, a few thousand in the bank, and a couple of old life insurance policies with your ex-wife."

Weber turned to him abruptly, his face crimson red. "You're a worthless wet-behind-the-ears piece of shit. Go fuck yourself!"

Adams tightened his lips as if to keep himself from saying what he was thinking, shook his head, turned, and left Weber fuming there in the hallway.

***

Weber didn't wait to get home to drink. He stopped at the nearest bar and wasted no time slamming shot after shot of cheap whiskey. Finally, the bartender, having been trained in such matters, refused to serve him anymore. She offered to call him a cab, having noticed he'd driven up and parked in front of the bar. She got the same treatment Sean Adams had received just an hour earlier.

***

Alina cringed when she heard the sound she'd come to fear: the tires crunching on the driveway gravel, the car pulling up to the house. Dinner was ready, and the table was set, but she never knew how bad he'd be at the end of any day. Always bad, just often worse. It took longer than usual for him to make it to the front door. This, she knew, was always a bad sign. She stood at the front door, about six feet back, waiting for the inevitable. The door knob turned, then nothing. It turned again, and Weber fell through the doorway and to the floor onto his knees. Alina reached to help him. The stench of whiskey and vomit made her gag.

"Get your…your…fuck…fucking h-hands off me, you…whore! Get off me!"

"Let me get you some water," Alina said.

Weber got to his feet and brushed off his knees. Alina approached him with a glass of cold water. Weber took the glass and launched it across the dinette table, smashing the dishes. Alina ran to the bedroom and locked the door.

Weber stumbled across the room and kicked the cheap door open, splintering the doorjamb into pieces. Alina clenched a pillow against her chest, preparing for the blows to come. Weber took two steps toward her, stumbled again, passed out and fell to the floor. Alina struggled to hoist him onto the bed, took off his shoes, and left the room. She fell asleep on the couch.

Around ten, the doorbell rang, and someone was pounding on the door. She looked at the clock, then turned and was shocked to see Weber awake and sitting in his favorite chair with the TV on as if nothing had happened. He stood.

"Who the fuck could that be?" He said as he walked to the door. He looked through the side window of the door. "You believe this? It's my fucking lawyer. What's he want?"

Alina stood well back away from Weber and away from the front door.

Weber opened the door. "What the fuck do you want?"

Sean smiled, took one step into the house, raised his right arm stiffly.

Weber took two steps back, tripped on the throw rug, falling onto the couch.

Adams promptly pumped three thirty-eight caliber bullets into Weber's forehead. The top of his head opened like a clam shell. *There!* Adams thought, seeing what he'd done to Weber. *That's for Alina.....and Susan Cranch.*

Alina screamed, putting both hands to her mouth in shock, and began sobbing uncontrollably. Sean Adams looked at her, then panic flamed through him. He dropped the gun, slowly backed up to the open front door, turned, and ran.

"OH MY GOD!" Alina screamed, "OH MY GOD OH MY GOD NO!"

That was what she screamed.

What she was saying to herself was, "THANK YOU THANK YOU! THANK YOU!"

She began hyperventilating, felt her knees weaken and sat at the dinette table staring at Weber. He suddenly slumped to the side, and his brain spilled out onto the couch, a bloody, gray gelatinous blob. She screamed again.

Alina sat paralyzed in thought and body. Finally, she found the strength to get up, go to the kitchen, and pour herself a glass of water. She held it with two hands trembling so severely that water splashed on her cheeks as she tried to sip. She knew she had to call someone, probably the police. She set the water down and, gaining some control over her emotions, walked over to the still-open front door. She stared at the gun on the floor near the threshold of the door. Then, slowly, she bent over and picked it up, first with two fingers on the barrel, then placed it firmly in the palm of her hand. It felt good; it was hot. She aimed it at Weber, put her finger on the trigger and contemplated firing one more shot into him for the mere satisfaction of it but chose not to. Instead, she took the gun and, still operating out of shock, walked to the bedroom and hid it between the mattress and box springs of their bed. As she picked up the cheap track phone, Weber had bought her to call the police from the top of the chest of drawers, a man called out through the open front door.

"Is anybody in here? Is anybody in here?"

Alina began to tremble again. Now what? She peeked out the bedroom doorway and saw two uniformed police officers standing just inside the front door with guns drawn.

Alina tried to shout, but nothing came out. She tried again, this time, she was able to barely get it out.

"I'm here! I'm here!" She said without moving.

"OK. Come out now," Deputy John Wilson, the first cop to enter, said. "nice and slow, with your hands in the air."

She complied.

She wore nothing but black leggings, a loose gray sweatshirt and black ankle socks. He swiftly took her by the left arm, swung her around and cuffed her. The other officer, Deputy Sam Irving, was staring at Weber on the couch. He was a rookie and had never seen anything even close to this. He suddenly turned away, bolted out the front door and vomited in the driveway.

A few minutes later, the Crime Scene Investigators (CSI) showed up. Detective Cindy Maddison headed up the team of four investigators. Cindy always chose to dress casually. There was no sense in wearing good clothes to a crime scene. She wore tight black jeans, a long-sleeve plaid shirt, and a black leather jacket. But she never gave up her three-inch heels, being just barely five-foot-one.

"Other than the obvious, what can you tell me?" she asked Deputy Wilson.

"Well, we got the call from dispatch just as we were driving down Shore Drive that a neighbor thought they'd heard gunshots from here."

"Kinda isolated, though, don't you think? How'd anyone hear anything?" she asked.

"When we got here, the front door was wide open. I guess if a gun was going off in here—and it sure looks that way—you'd be able to hear it all over West Shore."

"Makes sense," Maddison said.

Just then, one of the investigators called out from the bedroom.

"Cindy! You gotta see this," he said.

Cindy nonchalantly walked to the bedroom where her assistant was holding the queen-size mattress up, exposing the .38 revolver.

Cindy turned to Alina. "What's this?" she asked.

Alina said nothing, continuing her stare to the floor.

"OK. Grab an evidence bag, but photo it in place and dust it first," she said as she walked over to the bed. He complied. She was already wearing the requisite rubber gloves, so after they took several pictures of the gun and bed, she gingerly picked it up with her thumb and forefinger and handed it to the fingerprint specialist. "Three empty chambers," she noted.

Another of the investigators walked into the bedroom. In his glove-covered hand were three empty .38 caliber shells.

"Did you photo them where you found them?" She asked.

"Sure did," he replied.

"Dust 'em bag' em," Maddison said. She turned to Deputy Wilson. "Looks like an open and shut domestic violence case."

***

As they drove Alina to the sheriff's department, cuffed in the back seat of the deputy's car, she sat numb in catatonic silence.

"Anything you want to tell us?" Deputy Wilson, who was driving, asked.

She said nothing. She remembered what happened to people in Russia who were picked up by police in her town—they were never seen again. Best, she just stays silent.

"We're just trying to help," the other deputy, Sam Irving, said. "You know, the more you tell us, the better things will go for you."

Silence.

"Did you shoot him? I mean, it's pretty clear you shot him," Wilson said.

Silence.

"You know, if you're not going to deny it, maybe that means you did it."

Silence.

"Look," Wilson said, "There was nobody there but you and the stiff, so it had to be you."

"Lay off," Deputy Irving said.

Then she remembered a scene from a crime show she'd watched. She uttered one word.

"Lawyer."

"She say something?" Wilson asked.

She repeated it loud and clear. "Lawyer."

The deputies looked at each other. They drove on in silence.

# CHAPTER

# FIVE

## Annapolis

The server, 18-year-old Stephanie, approached Vince and Mary Beth at the dinner table.

"Can I get you all anything else?" she asked.

"More coffee would be nice," Vince said.

A minute later, Stephanie returned with a fresh pot and refilled their cups. "You sure I can't bring you anything else?" she asked.

"We're all good, thank you," Vince said.

"Vince," Mary Beth said, "she's wondering what kind of tip she's going to get for two cups of coffee."

"Don't worry, Mary Beth. I'll take care of it."

"You don't; I will."

Vince was flipping through the contents of the oil and grease-smeared manilla folder and pulled out the dock rental agreement.

"That's odd," he said.

"What's odd?"

"The slip rental agreement is for twenty years."

"Is that normal?" Mary Beth asked.

"Absolutely not. It's always one year at a time."

Vince scratched his head as he read the fine print.

"It's 'freely assignable.'"

"So?" Mary Beth asked.

"Have you ever been to Annapolis, Mary Beth?"

She smiled. "Vince, I spent a year doing an investigation at the Naval Academy here, remember?"

"No kidding."

"I told you that. Lived here a whole year."

"Did you like it?" Vince asked.

"No."

"Really?"

"I loved it."

Vince took a sip of coffee, then launched into his sales pitch. "Mary Beth, Annapolis is a mariner's paradise. Every service you could ever ask for. And the town is historic, beautiful..."

"What are you selling me, Vince?" She asked as if she didn't know.

"Well, look. This dock slip rental contract has years left on it."

"Yeah, so?"

"So, what if we were to buy the boat and keep her here instead of bringing her up to Ithaca? I mean, once we get her fixed up, we could stay on her on weekends, cruise the Chesapeake...."

"We?"

"What?"

"You said once *we* get her fixed up."

"Oh. Sorry. Right. I meant once *I* get her fixed up..."

Vince opened the manilla folder he had on his lap and pulled out the work orders from the last owner.

"Wow," Vince said. "The last guy put about a hundred grand repairs into this thing! Maybe I won't have to do much fixing up to do!"

Mary Beth chuckled softly, "Oh Vince, we're partners. We're partners in everything. Besides, I have better taste than you do. *We'll* fix her up. Let's go make a deal."

Vince was so excited he left Stephanie a fifty-dollar bill for a tip, and they left to go see Bill.

***

Bill sat patiently waiting on the back deck of his boat. His cellphone sounded; it was Rick Anders.

"Any luck, Bill?" He asked.

"They went to get something to eat and 'think about it.'"

'That's not good. You should've pushed it, Bill."

"Are you kidding? He's a fucking lawyer for chrissakes. You don't push a guy like that."

"Eh. You're right. Maybe lower the price?"

"Hm. Maybe. But that might make 'em suspicious, don't you think?"

"Fuck it. I'll leave it up to you. Good luck."

"Hey, they're pulling up now. Talk to you later."

Bill was an old hand at all this; he'd been selling boats for fifty years. He knew he had a deal the minute he saw Vince and Mary Beth get out of the car. They exited simultaneously, turned, and smiled at each other, then held hands as they walked down the dock. Bill checked to make sure he had a pen handy.

"So, what's the verdict, counselor?" Bill asked as the two came up the boarding stairs onto his boat's back deck.

Mary Beth and Vince sat.

"It's getting chilly," Mary Beth said, pulling her coat closed.

"We could go down into the cabin, if you'd like," Bill suggested.

"No, no. I'm fine," Mary Beth said, recalling the strong smell of diesel inside the old Matthews yacht.

Vince looked over at Mary Beth with his eyebrows slightly raised as if asking a silent question, and she gave a subtle nod to answer, "OK, yes."

"What's your best price?" Vince asked.

Bill sat back in his chair. "Hmm. Well, you got an offer?"

"Well…" Vince started.

"Twenty Thousand," Mary Beth blurted.

Vince gave a little chuckle, completely taken by surprise, then made a serious face. "Yes. Twenty." He looked at Mary Beth, loving her more by the second. She gave him a wink.

Bill would've taken ten cents to get that boat out of the marina. He stroked his chin and frowned. "Wow," he said. "Twenty grand, that's low. You could get more for that just by salvaging the mahogany she's made out of. Look, this is an estate, and I'd like to get the best I can for the kids," Bill said.

"Kids?" Mary Beth asked. "I saw the Coast Guard registration in the file under the name Victor Bennington. Was there no spouse?"

Bill took an audible deep breath. "No." His eyes darkened. "Look, I can go thirty."

Vince raised his index number to speak….

"We can do twenty," Mary Beth said.

"Yes," Vince confirmed. "Twenty."

"She's a tough one, isn't she," Bill said with a smirk.

"When she needs to be," Vince said, always proud of her.

Mary Beth smiled.

Bill stood. "I gotta make a call. I'll be right back."

He disappeared down into the cabin and called Rick.

"They're offering twenty grand, Rick."

"Fuck. Grab it. It's all going into our pockets anyway."

"OK"

"And for fuck sake, let's get that fucking boat out of here, and you know why."

"I do."

"Alright then. Close the deal," Rick insisted.

Bill reappeared on the back deck and sat. "I talked to the attorney for the estate. He's not pleased, but he approved the sale."

"Who's the attorney?" Mary Beth asked.

Bill wasn't ready for the question. There was no way in hell he'd tell her who the lawyer was they'd used for the Bennington estate so they could sell the boat; the same lawyer they'd used for every widower's boat. The only other lawyer's name that popped into his head was the U.S. attorney he had a run-in with a few years ago over a boat he'd sold that sank in the Potomac River in DC.

"Sandy Mitchell," Bill said, instantly regretting it as soon as he uttered it.

"I thought you said, '*he's* not pleased,'" Mary Beth said.

"Did I?" Bill said with a chuckle. "I meant 'she' of course."

Vince pulled the already-prepared bill of sale from the folder, crossed out $40,000.00, wrote in $20,000.00, initialed the change, and signed it. Bill took it from him, signed it, tore off the top copy, and kept the carbon copy.

Mary Beth pulled out her phone and looked up at Bill. "Do you want payment by Venmo, Paypal, or a direct transfer?"

Bill looked confused. "Can you write a check?"

"Of course," Mary Beth said. "Who do I make it out to?"

"Ah. You can just make it out to me. I'll disburse it. Make it out to William J. Brown."

She pulled out the checkbook on their joint account and looked at Vince. He nodded; she wrote the check and handed it to Bill.

The three stood up.

"Well, congratulations! You are now the owners of a magnificent classic yacht!" Bill said, reaching his hand to Vince.

They shook hands. Mary Beth kept hers buried deep in the pockets of her camel coat.

Bill reached into his shirt pocket and pulled out a business card.

"Here's the name of a great local captain, Randy Spaulding. He can cruise your boat up to Ithaca. Real experienced. He owns a 1930 Defoe Commuter yacht—the love of his life. I already talked to him, and he can start the trip on Saturday. Those engines are up and running smooth as can be. And look, I checked the charts, and it's a long cruise: you gotta go up the Chesapeake Bay across the Chesapeake-Delaware canal, down the Delaware Bay, around Cape May, at sea for two days up the entire New Jersey coast to New York harbor, up the Hudson, then through twenty-four locks to get to your lake. He says he'll do it for a grand."

Vince looked the card over. "Thanks, Bill. We'll think it over."

Bill frowned.

"Well, Thanks for everything, Bill. We're excited," Vince said.

Mary Beth took Vince's hand and gave it a gentle tug—the signal to leave.

They got into the Tesla and decided to take a tour of Old Annapolis. As soon as Vince pulled out of the marina parking lot, Beth spoke up.

"Vince, I know Sandy Mitchell."

"No kidding," Vince said, not concerned.

"She doesn't do estate work."

"Well, maybe this time..."

"No, Vince, she's one of the top federal prosecutors in the Northern Division of the United States District Court for the District of Maryland. Remember I told you I handled an investigation at the Naval Academy here in Annapolis?"

"Yeah."

"She was the U.S. Attorney on the case."

"Wow."

"Yeah, wow. Something doesn't ring well for me, Vince."

"Maybe she's retired," Vince said. "Maybe she's doing civil work now."

"Vince, she's 38; she's my age."

"Well, you're retired," Vince said, immediately regretting he'd said it.

"Oh come on, Vince. You know that's not true. And you know I had to leave the FBI because I was assigned as an undercover to protect you in that CCE case, and we fell in love."

"Best thing that ever happened to me," Vince said smugly.

"Which? That I saved your life, or we fell in love?"

"Exactly," Vince said.

They both sat smiling.

A minute later, Mary Beth piped up. "What was the last owner's name?"

"Victor Bennington," Vince said.

"Victor Bennington."

"Yup."

They drove on in silence. Mary Beth struggled to rein in her suspicions, but her years at the FBI just would not let her. After a visit to the Maritime Museum, a couple of marine supply stores, and a quick seafood dinner, they started the drive home with plans to head back the next weekend better dressed to be on the old boat, with ample cleaning supplies and food for the weekend.

An hour into the drive, Vince's cellphone sounded over the Tesla sound system, and his son Michael's name appeared on the screen. Vince tapped the green "accept" tab on the screen. Mike was at the DeMarco Law office in his maroon leather chair at his oversized mahogany desk, wearing a slim two-piece dark gray suit. Michael was pretty much a younger, more fit version of his father, Vince, and in practice, much more organized and always more prepared than his dad.

"Hey, Mike! How's everything?" Vince said in his happy-to-hear-his -son's-voice high pitch.

"Good Dad."

"Say hello to Mary Beth. We're on speakerphone."

"Of course you are," Michael said. "Hey, Mary Beth."

"Hi, Michael."

"You want to hear about the boat, Mike?"

"Sure, Dad, but maybe later."

Vince sensed concern in his son's voice. "What's the matter, son?"

"Nothing, Dad. But we got a call from a guy who says he's from the Russian Consulate in New York City looking for you."

Vince laughed. "The Russians are looking for me?"

Mary Beth raised an eyebrow.

Mike's tone was serious. "Yes, Dad, the Russian Consulate is looking for you. Do you want me to have them call your cell, or do you want me to give you the number?"

Vince and Mary Beth exchanged glances.

"Did you ask them what it was about?"

"Of course I did," Mike said almost tersely. "He said he'd only talk to you."

"Who's 'he'?" Vince asked.

"He said his name is Nick Volkov or something like that."

"No idea," Vince said.

"Yeah. Me neither," Mike said. "Could be a scam."

"Probably. Well, don't worry about it. If he calls again, I'll take it," Vince said.

The name sounded vaguely familiar to Mary Beth. She kept it to herself, sat back and tried to relax. Suddenly, the memory hit her with a jolt, and she sat up straight in her seat.

Vince noticed her sudden movement. "Mary Beth, are you OK?"

She took a breath, her mind spinning, and wasn't sure what, if anything, she could disclose to him.

Sandy...Nikolai...the Naval Academy...the boat that sank in the Potomac River in DC...Bill's odd mention of Sandy's name... She struggled to maintain her composure. She needed time to think—lots of time. Maybe it was nothing...

# CHAPTER SIX

### Ithaca

Gary Becker had reached the zenith of his legal career with a slam-dunk reelection, securing his third term as district attorney. In his mind, nothing surpassed the sacred duty he held. He was a relentless workhorse who routinely sacrificed sixty to seventy hours a week to his work, and often more when in the midst of trial. Appearances never swayed Gary; it was results he cared about. He was five-foot-six and solidly built, with unkempt, dirty-blonde hair, and he refused to wear a tie in the office. In court, he sported a tie but seldom pulled all the way up. At trial, he was disheveled, focused and unyielding in pursuit of guilty verdicts.

This would be District Attorney Gary Becker's first murder case in years; Ithaca was a safe college town. The few murders that did take place there were mostly drug deals gone wrong, lots of witnesses, lots of forensic evidence, an overworked public defender, quick plea bargains, confessions, and years apart. From what the sheriff told him, this seemed to be a straightforward domestic violence case—an argument that went too far and a handy gun— a bad combination. Becker had never liked the psychiatrist Weber. The few times he'd encountered him in court cases, he'd been rude to the point of abusive, short-tempered, suspicious, and dilatory in his work. He smelled of alcohol.

***

Detective Ken Ludlow had been at it a long time, having been a homicide detective with the NYPD for over thirty years. Other than being completely bald and proud of it, he was as physically fit as he was when he was forty—a big man, 6'3" tall, and 180 solid pounds. But he had gotten tired of the rat race, the endless flow of senseless murders in the big city, and decided to move back to his hometown, Ithaca, and take a job there. He considered it almost a retirement—a low-pressure job compared to what he was used to.

He sat in a metal folding chair in the windowless, concrete-block interview room at the sheriff's department, waiting for the deputy to bring her in. Wilson escorted Alina into the room, still in handcuffs and sat her in the metal chair across the green metal table from the Detective. She struggled with the handcuffs behind her back, trying to center herself in the chair, the cuffs clanking against it.

Ludlow took one look at her, turned to Wilson, and said, "Take those off!"

The deputy promptly complied and removed the handcuffs. Alina rubbed her reddened wrists.

Ludlow considered himself a quick judge of character. What he saw in Alina was a young, frail, confused, frightened, pale and thin girl— too pale and thin. To him, trademarks of a standard form of domestic abuse: isolation. Not that this would exonerate anyone from out-and-out murder, but he knew all too well that a person can only take so much abuse before they snap; then anything can happen, even murder. It's sometimes the only way to escape it, other than suicide.

Alina sat in the cold metal chair, self-hugging and trembling.

Ken looked up at the deputy. "Did you bring her coat?"

"I didn't see one," the young deputy replied. "Wilson brought her in like this."

"Well, get her a coat, and get it now!"

"Yes, sir."

The deputy left, closing the heavy steel door behind him. Alina looked and noticed there was no doorknob on the inside of the door.

Ken began. "He'll bring you something. Would you like something to drink? Soda, water."

"Coffee," she said, looking down at the table.

"OK. Coffee," Ken said, rising from his chair. He walked over to a small steel box mounted on the wall behind Alina and pressed the button. An immediate response came through the box so loud it made Alina grimace.

"Yeah?"

"We need a coffee in here," Ken said.

"You?"

"No!" Ken said, then turned to Alina. "cream and sugar?"

She nodded.

"Cream and sugar," Ken said.

"Gotcha."

A minute later, the same young deputy arrived at the door with a cup of coffee and a navy-blue men's large sweatshirt with "TOMPKINS COUNTY SHERIFF'S DEPARTMENT" printed across the front in reflective yellow. He handed it to Alina, who promptly slipped it on and handed Ken the coffee. Ken set it on the table in front of Alina, and she quickly picked it up and held the warm paper cup in both hands. The deputy took Alina in, and their eyes briefly met. He gave her a half-smile; she diverted her gaze to the table. He left, this time, making an effort to close the heavy steel door gently. Even in that quick encounter, there was something that touched this neophyte deputy's heart.

Ken had a brown folder in front of him with the first crime scene report in it. He left it closed.

He waited as Alina took a sip of her coffee, her hands still trembling.

Ken Ludlow had sat across steel tables in concrete interrogation rooms with hundreds of murder suspects at the NYPD. He'd questioned the hardest, dead-eyed, remorseless gang-bangers, hit men, cheated drug dealers, hysterical fanatical political zealots, drug addicts that killed for ten dollars, and countless people that had simply had enough of life and took it out on random strangers in the alleys, subway stations, and tenement housing projects. He was used to dealing with these people. This was different. His first murder case in Ithaca and Alina. He was used to coming across tough, impatient, demanding, and relentless. And if the "perp" had an attorney, he treated the attorney the same way, sometimes worse. This was new territory, and he couldn't conjure up any 'fire' in him. Rather, he felt protective of her.

"Alina."

"Yes."

"I'm told you asked for a lawyer?"

She nodded.

"Well, that's probably a good idea. Do you know who to call?"

She paused.

"I don't know anything."

"Well, you can probably get a court-appointed lawyer, but it'll take a while."

"Court appointed? I don't have any money."

At that, Ken realized she hadn't been read her Miranda rights, and for a second, it worried him, but since she hadn't been formally arrested yet, it wasn't required. Besides, she hadn't said anything self-incriminating.

"If you can't afford one, an attorney will be provided to you at no cost," Ken said, almost verbatim from Miranda.

"Is there someone you want to call?" He asked.

"Yes. My mother."

"OK. I'll get a phone for you."

"She is in Russia. All my family is in Russia. Is it OK if I call Russia?"

"Do you know how to make an international call?"

"Yes."

"OK. That's fine. I'll be right back."

Ken left the room and returned with a 1990s tan pushbutton telephone. He plugged it into the wall near the floor behind Alina and put it on the table in front of her. "I'll leave you alone to make your call. When you're done, just tap on the window in the door. OK?"

"OK. Thank you."

Ken left the room, and Alina picked up the phone and held it to her ear. Her mind spun; she froze. She couldn't remember anything— how to dial for an international call, not even her mother's phone number. She held the dial tone to her ear until it went to a fast, busy signal. She slowly replaced the receiver on the phone, her delicate hand trembling. Ken peeked around the corner through the wire mesh window of the door and saw Alina staring at the phone on the table in front of her. She slowly picked it up again and put it to her ear, and began dialing, cautiously, one number... pause, another number... pause, until she was done. To her surprise, the call went through and began ringing. It rang five times, ten times, twenty times, and she began to panic but then remembered there was only the one phone in the house on the wall in the kitchen, and they had no answering machine. She let it continue to ring until her mother finally answered, clearly out of breath from running to the phone. They spoke in Russian.

"Mama."

"Alina! Are you OK?"

"No, Mama."

Her mother took a deep breath and sat on the wooden stool next to the phone on the wall. Her chest tightened, and she struggled to take another breath. "Tell me, my sweetheart."

"Mama, I can't talk about this on the telephone, but I need an advokat."

"You need an advokat?" she asked, her alarm growing.

"Yes, Mama, a lawyer. I'm in trouble. This is all I can tell you."

"Where are you?"

Alina didn't know where she was. She began to panic, then looked down at the sweatshirt she was wearing. "Mama. Tompkins County Sheriff's Department," she said slowly.

Her mother scrambled to write it on the pad they kept on the wall by the phone with a pencil on a string. She wrote it phonetically in Russian.

"Mama, you know what number to call?"

"I do, Alina, I will call..."

The call suddenly disconnected.

"Mama! Mama!" She shrieked, "Mama!"

She set the phone down, crossed her arms on the table, dropped her head into her arms and sobbed.

***

While Alina was on the phone with her mother, Ken had retreated to the Sergeant's office and was on the phone with District Attorney Gary Becker.

"Sounds like a pretty open-and-shut case," Becker said.

That's a phrase Ken had stopped using his first year as a detective. There was no such thing in the law.

"What do we have from CSI?" Ken asked.

"Not much. Just a cursory description of the scene. Pretty gory. No forensics yet. Found a gun and shells. Did she make any admissions?"

"She lawyered up in the cruiser on the way over here," Ken said.

"I heard. That, by itself, makes me suspicious."

"Oh, come on, Gary. You know better than that," Ken admonished.

"She made no statements to Wilson at the scene and said nothing in the cruiser either," Gary said.

"Well, then we certainly can't hold her," Ken asserted.

There was a long pause.

"Gary?"

"Yeah. You're right. Not until we hear from the scene investigators. But keep her on the radar; keep an eye on her."

"Will do."

"OK, keep me posted, Ken."

"You too."

Ken hung up the phone and leaned back in the chair. She certainly couldn't go back to her home; it was a brutal, gory, murder crime scene, and sealed off. And the deputies dragged her in without a phone, purse, or even a coat.

He went back to the interrogation room with a small plastic box containing the Tri-Tech GSR kit. Alina sat stiffly in her chair, her eyes gazing off into the distance at nothing.

"Did you reach your mother?" Ken asked as he sat down across the table from her.

"Yes."

Ken waited for her to say something further, but nothing came. He waited a few seconds.

"OK," he said. "We're not going to hold you, Alina. You can go, but before you do, I'm going to swab your hands and clothes for gunpowder residue. Do I have your consent to do that?" He asked, fully knowing that consent wasn't necessary as the courts had determined that swabbing for gunpowder residue was "non-testimonial and minimally invasive."

She nodded, "Yes."

He opened the sealed GSR kit, removed the rubber gloves, put them on, then took out the swabs and swabbed her hands, fingers, between her fingers, and wrists, using a 5% solution of nitric acid. The swabs would be submitted to the crime lab to be subjected to Scanning Electron Microscopy.

She didn't flinch.

"OK. We're done here for now. Can you call someone to pick you up?" Ken asked.

She slowly looked up and into Ken's eyes. "I have no one."

"Your family?"

"Russia."

"OK. Let me see what I can do," Ken said. "By the way, are you hungry?"

"No, thank you."

"OK. I'll be right back," Ken said, leaving the room. He went out into the hallway and called Jason Valor, the social worker Ken and the other detectives who had worked with, with domestic violence victims. Alina may have been a murder suspect, but as far as Ken was concerned, she was still most likely a domestic violence victim.

Jason sat on the couch in the living room of his one-bedroom apartment, in plaid pajamas and a navy-blue bathrobe, with his laptop and papers on the coffee table before him, typing a report for the family court judge on a child neglect case. He had just turned twenty-eight, fit and trim, blonde hair and blue eyes.

"Hey, Jay, it's Ken."

"Hi Ken, what's up?" Jason asked, looking at his watch.

"Well, it's a little complicated, but I need your help."

"Ken, it's always complicated, and you will always have my help," Jason responded. "Tell me."

"Have you heard about Weber?"

"The shrink?"

"Yeah."

"I heard he got shot."

"Well, he was murdered. I've got his wife here."

"Why?"

"Well, she was the only one at the scene and, quite frankly, is the suspect."

"Is she charged yet?"

"No, but there's a high probability she will be."

"So, what do you need from me?"

"We can't hold her—no real evidence yet. I have to release her now."

"And?" Jason asked.

"She's got nowhere to go, nobody to even pick her up. She's a Russian immigrant, and her family's all in Russia."

"Is she a risk?"

"Naw, I don't think so. You mean dangerous?"

"Yeah."

"God, no. She's probably about twenty years old, petite, frail."

"Judging a book by its cover, huh?"

"I guess so," Ken admitted.

"Well, if anybody can do that right, it's you, Ken. Look, I'll be up in ten minutes."

"Thanks, kid."

"You're welcome, Old man."

Jason showed up at the sheriff's office twenty minutes later in jeans, a gray sweatshirt and sneakers, and Ken met him as he entered the front foyer. Meanwhile, Ken had moved Alina to his personal office at the station so she would be more comfortable.

"She's in my office," Ken said as he greeted Jason.

They buzzed through the security door, and Ken led him to his office. Alina was sitting upright on the old, worn, green faux leather couch and stood when Jason and Ken entered.

"Alina, this is Jason. He's a social worker, and he's going to help you find a place to stay for a while."

Alina looked Jason over, about 5'9, lean, soft blue eyes, and a shock of sandy-blonde hair. He reached out to her and shook her hand. He was surprised at how soft, cold, and almost limp her hand was, but realized that she was probably in a state of shock.

"It's nice to meet you, Alina. I wish it was under different circumstances," Jason said, truly meaning it.

"It's nice to meet you," she said, barely audible.

"I understand you need a place to stay for a while."

"I think so," she said.

"OK. I'm going to take you to the women's shelter. There will be a room for you; we can get you some clothes and a nice hot meal."

Ken piped in, "Yeah, I don't think she's had anything to eat." He glanced up at the clock on the wall; it was just past midnight. "You OK to transport, Jay?"

"Sure. That's fine."

"Keep an eye on her, Jason. She may be a murder suspect. Understand?"

"Of course."

Alina cringed at those words, realizing for the first time they truly believed she was a murderer.

Jason escorted Alina out to his car in the parking lot—a black, fairly new Jeep Cherokee--and opened the passenger side door for her. She pulled herself up into the seat; he went around the other side, got in and started the car. He looked over at Alina, seeing her in only a sweatshirt, and cranked the heat to full. He pulled his phone from his pocket and called the women's advocacy center.

Rachel answered the phone.

"Hey Rachel, it's Jason."

"How's my favorite social worker?" she asked.

"I'm good. You?"

"Oh, living the dream, you know."

"Boy, do I. Hey, listen. Did Detective Ludlow give you a call?" Jason asked.

"About what?"

"Alina Weber."

"No. Why?"

"Because I'm on my way to you with her now. She's got nowhere to go."

"Jason, he never called, and if he had, I would've told him that we're all filled up. We're actually beyond capacity."

"You're kidding?"

"I wish I was, Jason."

Jason tightened his lips and shook his head, feeling sadness in his heart to know there were so many victims at the shelter.

"I'm so sorry to hear that, Rachel."

"We can hardly keep up, Jason. Well, you take care."

Jason clicked off the call and continued driving, his heart heavy in his chest. He turned to Alina.

"You must be hungry," he said, trying to move on and being hungry himself.

She gave him a sheepish smile and nodded yes.

"Good! Let's go get something to eat."

He needed time to think.

"What do you like?" He asked her.

"Like?"

"Yes. What do you like to eat?"

"Ah. Anything. Anything is good."

Jason thought. Hmm. It's almost 1:00 a.m. Everything's probably closed. Maybe Taco Bell. Maybe McDonald's.

Then it hit him. Where's she going to stay? He noticed she didn't have a purse, a wallet, or anything with her at all, and he didn't dare leave her at a motel. She was a murder suspect, for chrissakes, and Detective Ludlow told him to keep an eye on her.

They got food at the McDonald's drive-through and headed to Jason's apartment.

By the time they got there, the weather had taken a turn for the worse. The temperature had dropped below thirty degrees, and it started to drizzle a mix of rain and wet snow.

"Stay put," he told Alina."

He got out of the Jeep, ran around the passenger side, escorted her out of the car, wrapped his arm around her to shield her, and they scampered to the stairs to his second-floor apartment.

"The weather sure changes fast around here," he told her as he reached into his coat pocket for the key.

"Yes," she said.

He opened the door, reached around the corner of the wall, flicked on the light and, with a wave of his hand, directed her in.

She stepped in cautiously and scanned the room with her big, hazel, bloodshot eyes. It was a standard one-bedroom basic apartment with white walls, a tan fabric couch, an oval throw-rug over a tan vinyl floor, two matching brown wood and vinyl armchairs, and a round dinette table with three maple chairs. A 50" flatscreen TV hung on the wall across from the couch. To the right of the TV was the door to the bedroom. Behind the couch, the room was open to the kitchenette.

Alina was completely drained. She made her way to the couch, sat down, then leaned over, putting her head on the arm of the couch and promptly fell asleep. Jason shook his head, seeing the awkward position Alina was in with her feet on the floor and collapsed sideways on the couch. He gently lifted her legs, placed them on the couch and put one of the small couch pillows under her head. He went into the bedroom, came out with a blanket and tenderly covered her. He slumped in the armchair, wondering just how many laws or rules of ethics he had violated by bringing her to his apartment. *What choice did I have?* he thought to himself. He was exhausted and headed to bed. He was asleep in ten minutes.

Around three in the morning, Alina woke from the grips of the tar of a nightmare with a start and sat up. The room was totally dark, thanks to the rubber-backed drapes in the room, and she had no idea where she was. Her mind was shrouded in dread, and she was struggling to remember if Weber's murder was a dream or reality. Finally, she could take it no more and let out an ear-shattering scream. Jason had only been able to sleep a few minutes on and off and was awake when he heard her. He bolted to his feet, turned on the light in his bedroom, threw his bathrobe on and sprinted to her side.

"It's OK, Alina! It's OK! You're OK!" He sat next to her, and she threw her arms around him and held him so tight he had to struggle to take a deep breath. The light from the open door to his bedroom cast a warm, subtle light in the room. She began hyperventilating; he held her close until she began to calm down.

"I'll get you a glass of water," he said.

"No. No. Stay with me, please. Stay with me," she begged.

They leaned back on the couch, and he held her until she fell asleep in his arms.

# CHAPTER SEVEN

The website for Russianbryde portrayed an expansive two-story modern glass and steel building surrounded by plush manicured grounds, palm trees, a pond with a flamingo floating in it, and a Rolls Royce and Bentley parked in front parallel to the building. In contrast to the darkly-tinted windows and black steel was the name, RUSSIANBRYDE, over the front entrance in polished brass letters.

No such building existed. Nor were there any offices of Russianbryde anywhere, unless the den in Ron Bryce's expansive townhouse in Miami Beach could be considered one.

Bryce sat on the balcony on the second floor, enjoying a Monte Christo Cuban cigar, sipping his Courvoisier cognac and enjoying the view of the ocean when his phone pinged.

He opened the call.

"We got trouble, Ron, real trouble!" Stan Reddinghouse said.

Bryce set his drink down on the glass table next to his chair and tapped the ashes from his cigar. "Calm down, Stan. What trouble?"

"We got one of ours murdered a client, that's what."

Bryce sat up straight and doused his cigar in the glass of cognac.

"Wait! Did you say one of ours killed a client?"

"Yeah. Another one! This is going to bring a lot of fucking heat on us, man."

"You sure it was one of ours?"

"I'm fucking sure. Her mother called me from Russia on the 'hotline emergency number' on our website, and it's all over the internet. Should I shut the site down?" Stan asked.

"Are you kidding? We took in a quarter-million just this month."

"It's gonna come back to us, Ron, it's gonna come back to us! And remember what the lawyer told us about wire fraud and money laundering, fucking life in prison."

"Shut the fuck up, man. Not over the phone, you idiot! Get up here."

"Fine. See you in an hour," Stan said, clicking off the phone. He got off his sixty-five-foot Ferretti yacht, ran down the dock, jumped into his Porsche 911 GTR and sped to Bryce's place.

***

Ron Bryce was the brains of the operation, but Reddinghouse was the tech guy. Neither could've pulled Russianbryde off without the other. Bryce came up with the idea, the theory that lonely, desperate, unattractive, rejected, angry men would pay almost anything to hook up with a young, beautiful girl and that impoverished, underprivileged, innocent and naive young women would easily fall victim to the promises of better lives—true or false. Bryce had done his research and knew the statistics, demographics, logistics, geography and finances, but he had no tech savvy whatsoever. He hired Stan Reddinghouse, who'd just graduated from Florida University with a master's degree in graphic art design whose specialty was designing commercial websites. He also graduated with over three hundred thousand dollars in student loans and other debt.

The two had met when Reddinghouse was a part-time waiter at the country club Bryce frequented in Miami Heights. Bryce had only attended one year of community college, but even before Russianbryde, Bryce was well-to-do from scamming hurricane and storm victims by showing up after a disaster in a pickup truck with a clipboard, pretending to be a contractor, taking cash deposits with the promise of returning with his "crew" and never showing up again. When that met with easy success, he enlisted dozens of coconspirators to run the scam all over Florida, taking fifty percent of the gross proceeds. It made him rich. He engaged Reddingtonhouse in a conversation, feeling him out for a possible addition to his construction scam business, but found it more enticing that he was a graphics art designer specializing in commercial interactive websites. That set the wheels turning in Bryce's head that led to Russianbryde.

Only four years ago, Bryce and Reddinghouse had put their heads together, and Regginghouse set up the website, using photos of models stolen from other websites and responding to inquiries himself posing as the young women. At the same time, they posted stolen pictures of male models and phony bios of "wealthy men" and geo-targeted the poorest regions of Russia with the website. With no real men or women on the site, the money came pouring in anyway, and the phony photos and bios were soon replaced with real men desperate to meet young women and young women hoping for a better life in the United States. As the money and numbers began to escalate beyond their control, Bryce hired Chad Whitcom, an algorithm specialist, computer scientist and professor who Reddinghouse knew from Tampa University in the computer sciences department. Whitcom was content with his sixty-thousand-dollar a year salary until he was approached by Bryce and Reddinghouse and offered a one hundred-thousand-dollar cash "sign-on bonus" and a quarter-million-dollar annual "tax-free" salary. Bryce figured that was a lot cheaper than trying to staff the site with hundreds of employees like the legit dating sites did.

It only took Whitcom two months to design a fool-proof, hands-off algorithm and to laser geo-target the site.

There was no filtering, no monitoring of the site, and certainly no vetting anyone; it was on full "auto-pilot," and no matter what "requirements" the men or women entered in the "desired traits questionnaire" the algorithm matched them randomly, so long as the woman was from an IP address in Russia and the man's credit card cleared. Once the upfront "retainer" from the "man seeking woman" of twenty-five thousand dollars cleared, the site left all further communications, agreements, travel, and relations—good or bad—up to the "paired" couple.

Now, they faced another problem that could unravel the whole operation.

"You know what to do," Bryce said.

"Again?" Reddinghouse asked.

"Again," Bryce said flatly. "Make your plans, and do it fast."

"'Volkov'?"

"She's Russian, isn't she?"

"Yeah."

"I'll have the cash for you tonight. You know the routine. Make your calls, and tell him to be ready."

"This has to stop," Reddinghouse said.

"It will when we do, and we're not going to stop. Now get the fuck out of here. I gotta get down to my boat."

"That old tub?" Reddinghouse said.

"Get out."

Bryce didn't joke about his yacht; it was the love of his life, the only thing he ever loved. One hundred feet long, one hundred tons, built in 1920—the former presidential yacht, *EMANCIPATION.* He'd gone so far as to buy the Bay Sound Marina in Annapolis, which had long since passed its better days. It was home to some of the oldest operating vintage and antique yachts in the country, and where there was one marine mechanic who knew his boat and its engines. It was a marina he'd known since he was a child and that held a special place in his heart.

He'd cruise her from her berth in Fort Lauderdale to his marina in Annapolis once a year for service and repairs. He didn't trust anyone else with his precious, rare, ancient yacht.

"Get out, Stan!"

# CHAPTER EIGHT

## Ithaca

Mary Beth and Vince got home around 1:00 a.m.

"Let's get to bed," Vince said as he unlocked the front door of the cottage on the lake they called home.

"Oh. No. Not until I take a shower, and you better, too."

Vince gave a puzzled look.

"Diesel!" Mary Beth said. "Diesel! We both smell like diesel, old boats and stinky cigars."

"I hadn't really noticed," Vince said truthfully.

"You must be kidding!"

"Well, maybe a little."

"And our car! It'll smell like that forever."

"Oh, aren't you overacting a little bit, just a little bit?"

She stopped dead in her tracks, turned and looked at Vince square in the eyes, slowly approached him, holding his gaze.

"You want to shower first?" Vince asked, a smile on his face.

She didn't answer. Instead, she moved close to him, gave him a lingering kiss on the lips, a little bite on his lower lip then took him by the hand to guide him to the shower.

"We'll go together. Wouldn't want to waste water, " Mary Beth said in a soft, low, inviting voice.

***

Sunday mornings were sacred to Mary Beth. It was the time she and Vince closed out the rest of the world, relaxed, and focused on each other. No television, no work, just easy conversation, reading, relaxing, cuddling and nurturing. The only exception was Vince's son, Michael. His calls would always be answered.

The cottage was a simple, cozy place—one story, vaulted ceilings, stone fireplace, two-bedroom, two bath, with a living room with wall-to-wall windows facing the lake. It was late October, and the view of West Hill with the leaves at peak color across the deep-blue lake was gorgeous and brought peace to the soul. An occasional sailboat sailing down the lake made it even more romantic. They sat together on the faded blue leather couch facing the view of the lake. Vince sat at the end with his feet on the matching ottoman; Mary Beth sat lengthwise on the couch at the other end, her feet in Vince's lap. His hand rested on them, occasionally caressing her feet as they both read their chosen books for the day. Mary Beth was reading Elsa Schiaparelli and Coco Chanel, a historic, romantic novel by her favorite author, Anna Maclean. Vince preferred non-fiction—pretty much anything to do with the sea, boats, ships or airplanes.

Vince's phone sounded.

"Unknown caller," Vince said.

"OK," Mary Beth said without looking up from her book.

He clicked off the call. He went back to his reading, one hand massaging Mary Beth's feet. Ten minutes later, his phone sounded again.

"Mike," Vince said.

"Ok," Mary Beth said. "Go ahead."

"Hi, Mike. Everything OK?"

"Maybe. Did you just get a call from a guy named Nikolai Volkov?"

"No, but I just got a call from an 'unknown caller.' Maybe that was him."

"Probably. He just called me here at the office looking for you again. He said it's urgent. You want to call him back?"

"Michael, it's Sunday morning. And what are you doing at the office on a Sunday?"

"Well, he said it was urgent. I mean, you know, you get a call from a guy who says he's from the Russian Consulate saying it's urgent—"

"The Russian Consulate? You kidding?"

"Well, that's what he told Miranda."

"OK OK."

"And why am I at the office on a Sunday? Well, you gave me a lot of cases, so..."

"I get it. OK, text me his number, and I'll call."

Mary Beth looked up. "Vince?"

"I'm sorry, Mary Beth, but this should be interesting."

"I bet," she said, no smile on her face. She put her book down. "Well, go ahead and call."

Volkov answered on the third ring.

Mary Beth sat upright.

"Mr. Vokov?"

"Mr. DeMarco?"

"Yes. My son told me there's an urgent matter?"

"Yes. Yes. Thank you for calling. It is urgent," Volkov said, in perfect English, no accent whatsoever.

Vince had expected Volkov to sound like Boris Badenough.

Volkov went on. "I received a call from the U. S. Embassy in Moscow that the mother of one of our citizens called and said she received a call from her daughter, a young girl, who is apparently in trouble."

*Wow*, Vince thought, international intrigue.

"And how can I help," Vince asked.

"We believe she has been arrested and is in Tompkins County."

"OK. Any idea for what?"

"Our intelligence tells us murder."

Vince's eyes widened. He was interested now. "What else can you tell me?"

"I'm afraid that's all I know, Mr. DeMarco," Volkov admitted.

"And how do you come to call me," Vince asked.

"I'm afraid I can't reveal my source, Mr. DeMarco. All I know is that I've been told you're the best criminal defense lawyer in that part of the country."

"Well, I'm flattered, but—"

"Mr. DeMarco, this has importance and implications beyond the obvious. Will you at least talk to her?"

Vince looked at Mary Beth. She rolled her eyes to the ceiling and shook her head. She already knew the answer.

"I'll talk to her," Vince said.

"And Mr. DeMarco, we will pay your retainer, whatever it is, no matter what," Volkov said.

"Well, that's not an issue, but thank you."

The phone line went dead.

"He never even gave me her name," Vince said.

"Well, it shouldn't be too hard, Vince. How many Russian nationals do you think have been arrested for murder in this town?" Mary Beth said.

"Ha. Not many, not many."

"Why don't you call Ken?" Mary Beth suggested.

He pulled up the "favorites" on his phone and tapped Ken's number. It rang three times.

"Vince!" Ken answered so loudly Mary Beth could hear him clearly from Vince's phone.

"Hey, Ken."

"What's up, Vince? I thought you were retired," Ken said.

"Semi! Semi-retired," Vince said.

"So what brings a semi-retired lawyer calling me on a Sunday?"

"You got a Russian national in custody?"

"Why you asking?"

"Mike might be representing her."

"Tell me her name, Vince, and I'll tell you."

Pause.

"I don't know her name," Vince admitted.

"Oh, that's interesting," Ken mused.

"Come on, Ken, tell me what you got."

"OK, Vince. I guess we go back long enough. Her name's Alina Weber, the psychiatrist's wife. You know about the murder, don't you?"

Vince was shocked. Not about the murder but that he had gotten out of the loop so quickly.

"Actually, no. I've been trying to stay out of these things lately. I hardly even read the news."

"Well, looks like Weber's wife shot him. A Russian girl, maybe twenty or so."

"You got her in custody, right?"

"Actually, no."

"Really?"

"Yeah. Not enough to hold her, Gary said let her go for now, but we're keeping an eye on her. She'll probably be charged as soon as we get the forensics back. Couple of days."

"Where is she?" Vince asked.

"The advocacy center, safe and sound," Ken said.

"That's good. Then I can enjoy my Sunday. I'll check in tomorrow."

"That's your call, Vince."

"Yup."

"How's Mary Beth?" Ken asked.

"She fine, and she's right here."

"What's she, thirty-seven now, and you're sixty—"

"Bye, Ken."

"Bye Vince, you lucky bast—"

*-Click-*

Vince set his phone down and turned to Mary Beth. "Ken says, 'Hi,'" Vince said.

"That's nice," she said, "but this all sounds fishy."

"Fishy?" Vince said.

Mary Beth was already back deep into her novel. She flipped a page.

***

After he bolted from Weber's house, his lawyer, Sean Adams, Esq., drove around town randomly and aimlessly. He was in full-blown panic, and despite the cold weather, he was drenched in sweat. What he had just done seemed both a million miles and years away, and as if it was still happening, all at the same time. Suddenly, he thought about the gun. *Oh my God! Where's the fucking gun? What did I do with the gun?* He had absolutely no recollection of anything that happened or anything he'd done after pulling the trigger. He patted his pockets and looked all around the car—nothing. *Maybe I threw it in the bushes.*

He'd had that gun for years, since his first year in practice. He was retained by a local drug dealer who just wanted to "keep him on retainer just in case." In other words, there was no case; the dealer just needed a legit front, and he picked Adams for three reasons: he was clean-cut, smart, and wet behind the ears. He gave Adams fifty-thousand dollars in cash and a loaded gun to "hold on to." In his need for money, trying to start his own practice, and buried in student debt, he chose to ignore the red flags, if not the unethical and illegal aspects of the transaction.

He lived in daily fear that the thug would call on him at any time, but that fear faded over the years, and that dealer disappeared. Sean didn't know or care where he went. The fifty grand was spent, and the gun gathered dust for years until he finally found this use for it. The only work he ever did for the drug dealer was to handle a single speeding ticket for him.

His next thought turned to Alina, the girl he'd done this for. He had to have her; he didn't care how; he had to have her. He would have her. He suddenly realized he was still wearing the rubber gloves. He pulled them off and tossed them into the back seat. With horror, he realized he was spattered with blood.

# CHAPTER NINE

Thanks to the scattered, puffy, fair-weather cumulus clouds, the sun rising in the east and peering through the wall of windows at the cottage didn't overwhelm Mary Beth and Vince as they sat at the dining room table in the open area having breakfast. Rather, it glistened off the lake's surface and made it sparkle like a million diamonds. Although Vince wasn't much of a breakfast guy, he couldn't resist Mary Beth's avocado toast and a cup of fruit salad with his coffee. They sat close to each other at the small round table overlooking the view. Vince had made a decision that he was reluctant to share with her, or rather, afraid of what her reaction would be. But, as Mary Beth always said, and Vince agreed, they were partners in everything.

*Here it goes,* Vince thought.

"Mary Beth."

"Yes, honey," she replied, setting her coffee down on the table, turning to him, giving him her undivided attention.

He took a deep breath.

"I'd like to keep the boat right where it is in Annapolis."

"I know," she said.

"You know?"

"Of course I know, Vince. I read you like a book."

"And?"

"So do I."

"Really? ...Really?"

"Yes. Really-really."

"Fantastic! I think we're going to love it."

"Me too, Vince."

"OK. So good. I'll call Bill and give him the good news.

Just then, Vince's phone sounded; it was Gary Becker, the district attorney.

"Where is she, Vince?!"

"Who?" And why are you yelling? Take a breath for chrissakes."

"Alina Weber. Where is she?"

"I have no idea, Gary."

"Deputy Ludlow told me she was at the advocacy center. They never saw her."

"You got a warrant for her arrest?"

"You bet I do, Vince. So where is she?"

"Gary, if I knew, I'd tell you. You're not somebody I'd mess around with. Let me do some digging. I haven't even met her yet. And wait! Why'd you call me?" Vince asked, wondering how Gary knew he'd been contacted to represent Alina.

"Oh, let's just say a little bird told me."

"A bird, huh? I'll give you a bird."

"Very funny, Vince. Good thing we've known each other for thirty years. I'll let that pass. Bye."

Gary clicked off the call.

Mary Beth finished her coffee. "They're pulling you back in, aren't they, Vince."

"I honestly have no idea what the hell is going on," Vince said in frustration. "I guess I better find this girl before they pick her up."

"How are you going to do that?"

"I'll start with Ludlow. He has to know something about it."

"Vince," Mary Beth said, "Are you sure you want to get mixed up in this case?"

"I don't know," Vince said, rubbing his forehead.

"You haven't been retained, Vince. No retainer agreement, no retainer."

"I know, it's just that...you know..."

Mary Beth cut him off. "Oh, I do," she said. "You're bored, and this intrigues you. It's hard to let go, isn't it?"

"I'm not bored, Mary Beth, but I think I owe it to this girl...or the Russian Consulate, I don't know. I guess I owe it to somebody to find her and at least talk to her."

"You owe it to yourself, Vince. I understand. Do what you have to do."

Vince stood from his chair, gave her a lingering kiss on the lips, and left the table to get dressed. He figured the best way to try to find out what this was about was to go see Detective Ludlow.

Mary Beth lingered over a second cup of coffee, enjoying the serene view out the window, wondering just how long their serenity was going to last. Vince wasted no time. He dressed casually in jeans, a tucked-in blue long-sleeve shirt, a navy blue pull-over sweater, and a brown leather jacket. He kissed Mary Beth and was out the door in a shot. She shook her head and focused on the view.

She waited until she heard his car leave the driveway, got up, grabbed her laptop from the bedroom, and started her research. She started with "Victor Bennington," the name Bill had given as the owner of the Matthews they had just purchased. The top hit was the obituary of Otto Bennington. She clicked on it and scrolled through the lengthy testimony filled with accolades and achievements, from lobbying for the first bill in Congress to protect the Atlantic Sturgeon to his "remarkable" golfing skills. In the last paragraph, it listed those who survived him, including his son, Victor Bennington and Victor's wife, Virginia. The next hit was Virginia's obituary—age 54, dated two years ago. She scanned it quickly, gasped, and read it again slowly. "Accidental drowning." After that, she searched for "Virginia Bennington." Not only did several news articles come up, but there were three different news clips from local news stations: one in Annapolis and two in Washington, DC. The articles were all nearly identical, reporting how she'd died. The earlier of the reports from DC said Virginia was attending a cocktail party onboard "a boat" as it cruised across the Chesapeake Bay from Annapolis. "All aboard agreed it was an unfortunate accident and likely a result of her intoxication, the pitch-black moonless night, and the inattention of the other partiers. No one realized she'd fallen over until they were almost to their destination." No charges were filed. The articles didn't specifically identify the boat, its owner, or its destination. The news clips were just as vague, with the exception that they showed photos of Virginia and Victor at different events as they described her death. They noted that a search was ongoing for her body. She continued to scroll, and just as she was wondering if the "estate" Bill had mentioned that owned the boat was the estate of Virginia Bennington, she found Victor's obituary—dated just one year ago. It was a single brief paragraph. There was no mention of his cause of death. He was 62. There were no surviving relatives mentioned—no children.

Next, Mary Beth searched the Maryland State online surrogate court records for Otto, Virginia, and Victor. Otto's was complete and professionally done. She pulled up his Will, which was short and to the point. He was predeceased by his wife, Ann, and left his entire estate to his son, Victor. There were no specific bequeaths, just "the res, residue and remainder of my estate to my beloved son, Victor Bennington," which translates to Victor gets it all, whatever it is. What he'd actually left Victor was massive debt and an IRS investigation.

She could find no records for Virginia. She did, however, find a handwritten petition for letters of administration filed recently, sloppily, and amateurishly done for Victor's estate. Since it was a petition for letters of administration, it meant Victor died without a will. Surprisingly to Mary Beth, because of the poor quality of the work, the petition indicated it was submitted by an attorney, Randolph Everts, Esq. It was filed as a "small estate" which meant his entire estate was worth less than fifty-thousand dollars, and listed no assets, nor did it identify any heirs. The petition sought to name William Brown, administrator of the estate.

Her next search was for Everts. There was no website for him or any firm he might belong to, but she was able to find his phone number and address through a site called "Lawyer Search." She found he was admitted to the Maryland Bar, suspended twice, and was currently in "probationary status."

She picked up her phone and called; it went immediately to voicemail, so she left a message, clicked off, set her phone on the table.

Everts didn't recognize the number that showed up on his phone and let it go to voicemail. If it was anything he wanted to deal with, he could decide when he listened to it. He closed his laptop, on which he had been watching YouTube videos, and listened to the voicemail message from Mary Beth. She'd simply said, "Hi, this is Mary Beth Allison. Please give me a call when you have a chance. Thank you."

Mary Beth continued to search for any more information on him—maybe a case he'd had or an appeal on the record. Nothing. Just then, her phone vibrated. It was the number she'd just called.

"Hello," she answered, "is this attorney Everts?"

"It is. I see I just missed your call; I was in a meeting. To whom do I have the pleasure of speaking," he asked, already taken with her soft voice.

"Well, first of all, thank you for taking my call. I know you must be busy."

"Oh, you have no idea," he said, fully knowing he had only one active case going, a client with his third drunk-driving charge. He leaned back in his gray metal and green vinyl chair behind his matching worn gray metal desk. He was short, stocky, with gray male pattern baldness, wearing blue jeans, tattered at the hems, loafers, and a brown polo shirt with a ketchup stain on the front.

"What can I do for you?" he asked.

"My partner and I bought a boat called *MY CHOICE* in Annapolis, and I understand you handled the estate for the deceased owner. I figured it'd be a good idea to speak to you directly so we can be sure we get clear title from the estate administrator."

Everts swallowed hard.

"How, how did you get my number? I mean, how did you know––"

"You're listed as the attorney of record on the Bennington estate with the surrogate's court," she said, cutting him off. Already, this call didn't feel right to her.

"Did you say you bought it, or you're going to buy it?" he asked.

"Bought it."

He took a deep breath, leaned forward, set his phone on the desk, put his elbows on the desk, and his head in his hands. He needed time to think.

"Hello? Are you still there?" she asked.

He slowly picked up his phone and sat up in his chair.

"Look, I'll need to pull the file, and I'm awfully busy, lots of deadlines. I got your number here in my phone. I'll get back to you soon as I can. OK?"

Mary Beth's FBI training was on fire.

"Sure, that's fine. Thank you," she said, clicking off the call.

He slowly set his phone back on the desk, staring into space. *Why did I ever let myself get mixed up in this?*

# CHAPTER TEN

**Two Years Earlier**

**Annapolis**

Victor bought the boat almost on a whim. He sat at his small desk in the glass cubicle at his job at the Subaru car dealership and was flipping through a copy of Boating Magazine someone had left there as he waited and hoped for someone to come in looking to buy a car. Victor was tall, all of six feet two inches, lean, with a full head of brilliant, trimmed white hair, and at work, always dressed impeccably in a dark suit, crisp white shirt, and polished shoes. One thing Victor was good at was selling. He'd always been the top salesman at the dealership and enjoyed what he did. Now, being just a few months from retirement and after working full-time since age eighteen, the thought of impending idleness panicked him. He was never a golf player, never had children, and had enjoyed a long marriage with his high school sweetheart, Virginia. But even that had lost much of its fulfillment for him. It seemed they didn't have much to talk about anymore, and they just seemed to coexist at this point. Retirement would only make it worse.

He scanned through the articles, all showing sunny days, blue waters, happy faces, and describing nothing but excitement, joy, and adventure. Every boat featured was described in superlative terms, "the best, seaworthy, family-perfect, best entertainment, safe, cozy, comfortable, fast, economical!" By the time he got to the classified ads in the back of the magazine, he was in love with the idea of captaining his own boat. He convinced himself that would bring happiness back into his life, and he pictured himself and his wife Virginia on board with ear-to-ear grins, toasting each other with martinis on deck as they cruised on serene waters under blue skies.

He perused the ads with something between interest and excitement. The prices ranged from fifty-thousand to ten million. At the low end of the spectrum were twenty-foot speed boats; at the top were hundred-foot mega yachts. None of those would do. He scanned the "prior-owned yachts" page. Even there, anything nearing the size of the gorgeous cruisers in the articles was hardly less than a million dollars. At the very bottom of the page, in the last column, was a picture of a fifty-three-foot Matthews cabin cruiser at sea crashing through large cresting waves; it was breathtaking. So was the price: "$60,000 or best offer."

He called on it and went to see it straight from work without even telling his wife, Virginia. The boat broker was ready for him, financing included; just the buyer's name blank on the bill of sale and finance agreement.

He toured the boat and was thrilled at how beautiful and big it was. There was a galley complete with dishwasher, accommodations in three cabins for six to sleep, a luxurious salon with wall-to-wall carpet, and two heads with showers. Having never owned a boat in his life, he had no interest in the engines, systems, fuel consumption or cost of maintenance. Being used to bartering in his job as a car salesman, he was comfortable dickering the price and offered fifty thousand dollars if he could get financing. The broker surprised him, saying financing was available for him, and all he had to do was make a five-hundred-dollar down payment; installments were five-hundred-fifty a month. Victor signed on the spot, and the broker gave him a receipt for his check, copies of the bill of sale, finance agreement, and a folder of photos of the boat in her earlier glory days, including the one that appeared in the magazine.

***

On the way home, Victor rehearsed his sales pitch to his wife. Selling is what he had done for the last twenty years, and he was confident he could sell this to her.

***

Virginia had retired from her job as an elementary school teacher a year earlier and spent her days now leisurely enjoying doing crafts she'd perfected in all those years teaching them to the children, cooking more, reading more, and generally enjoying being home. She had dinner ready and the table set for two, anxious for her husband to get home and spend a peaceful evening with him. She was sitting on the couch in her pajamas, slippers and robe, reading, when Victor came in the front door, took off his jacket and hung it on the coat rack. She put her book down on the coffee table, got up and greeted him with a hug and kiss on the cheek.

"What's that?" she asked, looking at the folder in his hand.

"It's a surprise," he said.

"A surprise?" she became curious and smiled. "What is it?"

"Come sit down," he said, taking her by the hand and leading her back to the couch. He opened the folder, showed her the photos first, and let her scan through them. The first picture was the one that was in the ad in Boating Magazine.

"Oh. Pretty," she said.

"Look at these," he said, pulling the interior pictures from the pile.

"Wow! Looks like a palace," she said.

"Whose boat is it?" she asked.

He smiled and paused.

"She's ours!"

"What!?"

"She's ours, Ginnie."

Virginia took the pictures and leaned back on the couch, slowly looking over each one.

"I don't understand, Vic," she said soberly.

"It's for us, Ginnie."

"You want to buy it?" she asked.

"Ginnie, I already did."

She laughed. "That's hilarious, Vic. We could never afford anything like that."

"No, we can! We can. Guess how much it cost?"

"Wait. You're serious. You already bought this?"

He nodded. "Guess."

"This is ridiculous," she said, shaking her head, still not believing him, "OK, I guess a million dollars."

"Nope. Fifty grand, and it's only five hundred bucks a month. That's less than your car."

"Really?" she said, sitting up. "All that for only fifty-thousand?"

"Yup. It'll be like a second home and on the water!"

"But Vic, we don't know anything about boats."

"What's to know?" he said. "It floats; it goes."

"Well, I think we can handle those payments if we cut back on other things," Virginia said.

"Ginnie, you're going to love it! Besides, we've been living right here in Annapolis for twenty years. Annapolis! The yachting center of the world!"

"Calm down. I'll trust you, Vic. Let's eat."

***

It was a horrible fight, something they'd never experienced before in their forty-year marriage. Sure, they'd had arguments, but nothing like this. The bills just kept coming and coming, and the money they'd saved for years for the Mediterranean cruise was all but exhausted. It was a dream Virginia had had since she was in college—a whole month cruising and something Victor had promised her. The brochure she kept and read over and over showed pictures of Genoa, Florence, Naples, Porto Fino, Sorrento, Capri, and Venice; all the scheduled stops on her dream cruise. She had never even been outside the United States. That dream was all that kept her going day to day.

Then, the bill for overhauling the two diesel engines showed up in the mail. She opened it, the blood drained from her face, and she felt her knees start to buckle. She made her way to the kitchen table and sat with a thump. She had no warning it was coming, and not only would it completely exhaust their savings, but she would have to put thousands on their home equity line of credit to pay the balance. She looked the bill over. It was pages and pages of parts; everything from sixteen cylinders at $3,760 each, boxes of bolts, belts, gaskets, oil lines, filters, rebuilt alternators, gallons of oil and lists of items she couldn't understand, and a single line: "labor/$75 hr. 162 hours: $12,150. Total tab: $86,412." She bolted from her seat, dropped the pack of bills on the floor, ran to the kitchen sink and vomited. When she recovered, she picked up the bill from the floor, went into the living room and sat on the couch in darkness, the invoice on her lap and waited for Victor to get home. She would give him an ultimatum; she would demand he make a choice.

An hour later, she heard the front door open. Victor was home. She didn't move.

'Ginnie?" he called out. "Virginia! Ginnie, I'm home," he said, walking from the foyer into the dark living room. "Ginni? What the heck? You OK?"

She didn't respond; she just held out the thick packet of invoices with the total on the cover sheet.

He took it from her trembling hand, reached over, turned on the lamp on the couch side table and looked at the cover sheet.

"Shit," he said. "I should've told you about this."

"You knew?" she asked, "YOU KNEW?" she screamed.

"Ginnie, Ginnie, calm down, calm down! I can explain!"

"Vic, you've ruined us!"

"No no no!" he said, sitting down on the couch next to her. "It'll all be worth it."

"Worth it? Worth what? Putting us in bankruptcy? I won't even bring up the cruise we've been planning for years! All the money we saved just for that gone! It's all gone, Vic! All our savings, all of it, and for what, that goddamned boat of yours? That fucking Goddamned boat!?"

"Ginnie, you'll see! It's beautiful, and you're going to love it, we're going to love it, and I didn't have a choice. The engines were shot and wouldn't run."

She was exhausted—physically, mentally and emotionally and slumped back into the couch.

"So, here it is, Vic. It's either me or that boat."

"What? What are you talking about, Ginnie?"

"it's that simple. It's me or the boat. You choose," she said in exasperation.

"Oh, that's just ridiculous," Victor said. "I did this for us! I did it for you! You'll see! Please, Ginnie, please. You can't say something like that after forty years of marriage."

Virginia began to cry, partly out of sadness, partly out of rage. She got up, ran upstairs, locked herself in their bedroom and took two Xanax tabs. She fell into a fitful sleep in ten minutes. Victor poured himself a glass of gin, plunked himself into his easy chair and put on the TV. He knew this was going to be an ongoing problem for him, but he loved his wife and wanted her to understand.

***

Two weeks had passed since that bill showed up, and Virginia had fallen into a somber, numb, subtle mode of passive aggression with her husband. As far as she was concerned, he had betrayed her, and things in the marriage would never be the same again. For the first time in years, she felt the deep pangs of regret for never having had children. At least she'd have someone to turn to, someone to find joy with.

***

Friday nights were the Bay Sound marina committee meetings. The owners of the twenty-four yachts docked at the marina comprised the committee, and gathered in a large room in the back of Rick's shop adjacent to the marina parking lot. Bill Brown was the chairman and had been for thirty years. Each member of the committee was entitled to a vote in all marina matters. Victor was attending his first meeting and was excited to be a part of it. Bill called the meeting to order.

"Good evening, gentlemen," Bill said.

They all responded one at a time, around the room. "Good evening, chairmen."

*Wow,* Victor thought. *That's just a little weird.* But when it came to his turn, he said, "Good evening, chairman."

"Gentlemen, we have a new member with us tonight. Please welcome Victor Bennington. He is the new owner of the fifty-three-foot Matthews, *AVENGER*."

They all turned, smiled at Victor and applauded. Victor nodded around the room to them.

"She's quite the classic yacht," Rick said.

"Thanks, Rick," Victor said, knowing Rick well from all the work he'd done on the boat.

"Rick keeps us all running smooth," Bill said.

Nods around the room.

"Victor, tell us about how things are going with *AVENGER*," Bill said.

"Uh. Where to start. Well, I've never been so excited about anything in my life. Rick has done an amazing job rebuilding the engines. I can't wait to get her out to the Chesapeake and maybe down to Cape May," Victor said.

"Ah, Cape May!" Tom Reynolds said. "That's a great cruise. You'll love Cape May. That's your gateway to the Atlantic. From there, you can get anywhere in the world!"

This excited Victor. Reynolds owned a sixty-foot 1950 Trumpy Yacht in pristine condition on the pier to the south of Victor's pier.

"You do that trip?" Victor asked, a smile on his face.

"More times than I can count," Reynolds said. "And that Matthews of yours can sure handle it."

"Wow," Victor said.

"And how's that little lady of yours," Bill asked.

The room went quiet.

"Uh. Well, she's fine, but, you know, I'm sure you all know when the wife gets the bills...well, you know," Vic said, looking down into his lap.

John Miller, the owner of a 1929 seventy-five Elco yacht, piped up. "Oh, we all sure know about that!" he said.

Chuckles around the room.

"She even told me I had to make a choice; it's her or the boat," Victor said, expecting chuckles.

Instead, there was an odd, sudden silence and glances around the room. Finally, Miller spoke up.

"Oh, we can help with that," Miller said.

"I wish!" Vic said.

"Oh, no. We really can. We're all used to this," he said with a big smile on his face.

Nods around the room.

"Sure," Rick said. "Let's do a local dinner cruise for her so she can appreciate what a great thing having that Matthews is."

"Holdings, you bring the blue crabs, scallops, and mussels," Bill ordered, then turned to Victor. "John Holdings owns the Blue Crab Shack downtown. He's got the 1930 sixty-foot Frank Anderson, just three docks down from your Matthews, Vic,"

"Yeah. That's a beauty," Victor said.

"Good!" Bill said. "We'll all bring something. Rick, you're in charge of the bar."

"My pleasure," Rick responded.

"This'll turn her around, Vic," Miller said.

Nods around the room.

Bill then moved the meeting. They covered topics including parking issues, diesel leaks in the water, garbage collection, curfews, whose boat was being repaired for what, and committee assignments for various general marina upkeep.

As Victor sat observing the proceedings, it suddenly struck him: everyone there was a male in his sixties, and there were no women in the group. In fact, he couldn't remember ever seeing any women in the marina. He let the thought pass.

***

When Victor told Virginia about the cruise that was "in her honor," she responded with complete indifference. She wouldn't agree, she wouldn't object, she'd just go along.

# CHAPTER ELEVEN

## Present Time
## Ithaca

Jason had slept fitfully on and off, but Alina slept motionless in a death sleep in his arms on the couch. He wasn't sure what kind of trouble this would cause for him, only that he was probably in trouble. The rules of ethics he'd memorized and knew could make it forbidden to engage in any romantic relationship with a client and certainly forbade any sexual relationship. So far, he knew he hadn't violated those rules, although her warmth, softness and vulnerability melted his heart. But most importantly, the primary ethical rule was that he do everything possible for the benefit of his client's health, safety and mental well-being. That, he truly believed he had done, was doing, and would continue to do for Alina, whom he considered his client from the moment Detective Ludlow had handed her over to him.

"Alina," he said, gently trying to pull from her. "Alina, are you OK?"

She awoke slowly. "Oh. Ah, yes. I'm OK, thank you."

She sat up, adjusted the Tompkins County Sheriff's Office sweatshirt she'd worn all night, looked all around the apartment, then at Jason, and realizing where she was, became alarmed. Then, the realization that what had happened the night before wasn't a nightmare but her reality. She gasped, covered her mouth with both hands and looked at Jason with terror and panic in her eyes. At first, she recoiled from Jason, then threw herself into his arms and sobbed uncontrollably. He did not embrace her but patted her on the back as if consoling a child.

His phone sounded. He knew by the number on the screen it was the Sheriff's Department. He wanted to answer—he knew he should answer, but Alina was in hysterics. He tried, ever so gently, to extricate himself from her but realized that until she could calm herself, he was in a vise grip.

The phone sounded again—the same number.

Then again. He let it go to voicemail both times.

Finally, Alina went limp in his arms.

"Alina, we have to go," he said, tapping her on her shoulder. She sat up.

"Where?" She asked.

"Well, first, I'm taking you to the advocacy center to get you fresh clothes, a good breakfast, and things like a toothbrush, comb, and hygiene; they have a kit you can have that has all that kind of stuff in it. Then, I have to take you to the sheriff's department."

She nodded in silence.

Jason played the first voicemail; it was from Rachael at the advocacy center, telling him they had an opening and a room for Alina. He didn't bother with the second voicemail; he knew it was the sheriff's department.

They got in Jason's Jeep and drove. Jason called the center en route. Rachael answered.

"Hey, Rachael," Jason said in an upbeat tone.

"Hey, Jay. You on your way?"

"We are, but she won't be staying with you, at least not now. Can you have some clothes and the usual stuff ready for her?"

"Sure. What size?"

"What?"

"I said, what size clothes, and what does she need?"

"Well, she needs pretty much everything," he said, turning and looking at Alina.

"OK," Rachael said patiently, "What *sizes*?"

"I don't know. She's pretty thin and--"

"Just get her here; we'll figure it out. Does she need medical care?"

"No, not that I can tell."

"Counseling?" Rachael asked, suddenly remembering Jason was a therapist. "Never mind. Just come in."

When they got there, Rachael guided Alina to the shower. Alina took her time, enjoying the calming hot water. She then took Alina to a room in the back of the center where they kept a large supply of donated clothes, both clean, used, and brand new, and gave her a choice of clothes that she thought might fit her. She picked a pair of slightly worn blue jeans, a white button-down shirt, a black jacket, and one-inch heel black pumps. Rachael carefully placed Alina's old clothes in a plastic bag as she knew there was a criminal investigation going on, and she did not want to be accused of destroying possible evidence. Rachel and Alina rejoined Jason, and she handed the bag to him.

"These are her old clothes, Jay. Be sure to give them to the police."

***

Vince pulled into the Sheriff's Department parking lot. He always preferred in-person contact, whether with a client, judge, cop or just about anyone in his life. He could get more information from a person by their posture, hand gestures, facial expressions, and glances around the room, and when their eyes met, than he could by the words they spoke. And so, he didn't call Detective Ken Ludlow to ask about Alina Weber; he would wait until they were face-to-face. Besides, he knew him so well he could read him like a book before Ken said a word.

It wasn't hard at all. Ken had seen Vince's Tesla pull up in front of the office through his office window and was waiting for him in the entrance lobby—his arms crossed tightly in front of him, his lips tight, his brows furrowed.

"My office," Ken said flatly as soon as Vince entered the foyer. Ken turned on his heel, walked to the solid steel door that separated the foyer from the internal offices, opened it, and pointed. Vince entered; Ken followed. No words were spoken until Ken sat behind his black metal desk, and Vince sat in the single black metal and vinyl chair in front of him.

Vince opened. "How you been, Ken?"

Ken softened his stance, leaned back in his chair, slightly dropping his detective outer shell, and allowed his friendship with Vince to slip through—just a little.

"Tired, Vince. Gettin' tired," Ken admitted.

"You work too hard, my friend," Vince said.

Ken scoffed. "Christ, Vince, I've never seen it this bad in this town."

"The crime?"

"The death," Ken said. "Six overdoses this week alone, three dead. Kids. Just kids."

Vince suddenly saw the sheer exhaustion in Ken's face and slumped shoulders. He took a deep breath.

"One of the kids that survived–only sixteen years old–told me the pills sell on the street for fifty bucks each. They think they're buying Percocet or Hydrocodone; they're getting pills made out of plaster and fentanyl. Pure poison. The ones that don't die are so addicted that they have no choice but to commit crimes to support a $ 500-a-day habit. It's a fucking spiral, Vince, and I don't see an end to it. We arrest these kids for possession, and they're back out on the street in three days. My deputies can't keep up with it."

Vince looked down into his lap and let Ken's words settle in. Vince was well aware of the epidemic killing teens and even children. That was why he never defended drug dealers.

"Maybe what we need is more social workers and fewer cops," Vince said.

"For the dealers?" Ken said, sitting up in his chair.

"Of course not," Vince said. "You know what I mean, Ken. For the addicts, especially the kids that get hooked on this shit."

"No, Vince. Even a few days up here in the jail should teach them a lesson."

"Sure. Maybe for stealing a candy bar from the store or getting into a fistfight on the street, but no three-day or three-month jail stint is going to be able to beat an opioid addiction," Vince said.

"Speaking of social workers, you know Jason."

"Sure, clinical social worker here."

"He took Alina Weber to the women's shelter last night, or I thought he did."

"Oh, I know. I got a screaming call from Gary, the DA, that she never showed up there. Did you call Jason?"

"I left three voicemails," Ken said, shaking his head. "I'll have to pick Alina Weber up on a warrant. I just hope I don't have to arrest Jason too."

Vince gave a closed-lip smile. "Naw. You can trust that kid."

Just then, Ken's intercom buzzed.

"Jason on line one," the dispatch deputy said.

"See?" Vince said.

Ken smirked, picked up the phone and hit the 'line one' button.

"Where is she, kid?... Better be...OK...bye."

Ken hung up the phone. "They're on their way. Wanna wait?" He asked Vince.

"Of course I do."

"Just to be clear, Vince, I'm going to arrest her the second she walks in the door."

"I figured," Vince said. "You let me talk to her first?"

"Five minutes, Vince."

"Gee, thanks."

"You can use the file room. There's a table and chairs in there."

"Yeah. I've only used it. What, a thousand times, Ken?"

"Smart-ass."

"What do you have on her?" Ken.

"Well, let's see," Ken said as he opened the file on his desk in front of him. He flipped through five pages. "She was alone with Weber when my guys showed up, nobody else anywhere around. Found the gun hidden under the mattress in the bedroom; pretty sure her fingerprints were all over it, no other prints. Three slugs in Weber's forehead. Gee, besides that, what else would I need?"

"For probable cause, that's plenty to arrest her," Vince had to admit. "Conviction, that's another story, and 'pretty sure' on the fingerprints doesn't mean anything."

Ken chuckled. "Oh, Vince, my old buddy. I gotta love ya."

"Blow-back on her hands, clothes?" Vince asked. "Gunpowder?"

Ken's intercom buzzed. "They're coming in the front door," the deputy announced.

"Come on, Vince. Let's go."

They left Ken's office; Ken escorted Vince to the file room and left him there. He went out to the entrance foyer and escorted Jason and Alina to the booking room where Detective Wilson was waiting to fingerprint, photo Alina and take a DNA swab from her inner cheek.

"What about him?" Wilson asked Ken, tilting his head toward Jason.

"He can stay for now if she wants him to."

Alina nodded her head emphatically.

Jason handed the clear plastic bag containing Alina's old clothes to Wilson.

"Go ahead and process her. I gotta make a call," Ken said. He returned to his office and called the District Attorney, Gary Becker.

"Yeah. She's being processed now, Gary," Ken told him over the phone.

"OK, good."

"Second-degree murder?" Ken asked, wanting to know what charge to formally arrest her for.

"Nope."

"No?"

"No, Ken. I don't want to over charge her and then have to back down, and I won't charge the top charge until we at least can tie the actual shooting to her."

"I don't get it, Gary. What else do we need?"

"GSR, Ken. We don't have gunshot residue on her yet. Nothing back from the lab on that. Until I have positive results that there was gunshot residue on her hands, I can't directly put that gun in her hand when it was shot. And since there were three shots fired, and it was an older revolver, there would be a ton of gunshot residue on her if she fired that gun."

"I just got the clothes she was wearing when we picked her up at the crime scene," Ken said.

"OK, good, so those go to the lab for blood spatter and GSR analysis. That'll take a while."

"So?" Ken asked, understanding but frustrated nonetheless.

"Book her on criminal possession of a firearm."

"You fucking kidding me?" Ken asked incredulously. "An E felony? One to four years and a fine? Damn. First-time offender, she'd probably get probation."

"Yeah, maybe, but I'm serious. Look, we'll make the judge understand the circumstances and that it was a brutal murder, and we'll ask for a million bail, two million bond. At least that way, we'll have her when the lab reports come in, and we can book her for murder. She'll never be able to post that kind of bail."

Ken understood; he didn't like it, but he understood.

"We gotta put that gun in her hand for a murder charge, Ken. Get her processed, charge her with 265.01, a gun charge and transport her down to City Court for arraignment. I've already called the clerk so they know she's coming. They're not in session, so you can get her arraigned and back to jail quick. I'll see you there."

Ken hung up with Becker and went into the booking room where Deputy Wilson had finished processing Alina. She sat handcuffed to a wooden bench. Ken removed the handcuffs from the large metal ring on the bench and led her to the file room where Vince was waiting, keeping her still handcuffed.

"I'll let you two talk for a couple minutes. The judge is waiting in court to arraign her. I'll be back, and we'll transport her down there. You can meet us there, Vince. We're charging her with criminal possession of a firearm, Penal Law 265.01. Here's the charge."

He handed Vince the paperwork, left, and closed the door behind him.

Vince gave it a quick glance. He was familiar with the charge and the possible sentence, and he was well aware that this was only the opening salvo from the DA.

"OK, Alina. So, in a few minutes, they're going to take you in the deputy's car down to City Court. I'll meet you there. It'll be a brief appearance. The prosecutor, Gary Becker, the DA, will present the charge of criminal possession of a firearm against you. I will enter a plea of not guilty on your behalf, and the judge will set bail or bond."

"Bail or bond?" Alina asked, barely able to follow what Vince was telling her.

"Yes, an amount of money that you would have to pay into the court so the judge would release you until trial."

"How much?"

"Oh, I don't know. You don't have a criminal record, so ordinarily, it might be five or ten thousand dollars or even release without bail, but there's a murder here, so it will most likely be more like fifty thousand, something like that. Regardless, I'm going to argue for release without bail, maybe supervised release, and we'll see what the judge decides."

Alina barely followed what Vince was telling her.

"Will you ask Jason to come too?" She asked.

"No need to ask; he'll be there, I'm sure."

***

When Vince got to the courtroom, DA Gary Becker was the only other person there. He stood at the prosecution table—the one closer to the jury box—and was carefully reviewing the documents there. Vince approached him, Gary turned, and they shook hands.

"Criminal possession of a firearm," Vince said matter-of-factly.

Gary nodded. "For now."

He handed Vince a manilla folder with the tab labeled *People v. Alina Weber.*

Vince took it, walked over to the defense table, sat and opened it. Just then, his son Michael came into the courtroom and took a seat at the defense table with his father. Vince slid the folder over to Mike, who opened it and began perusing the paperwork.

"Criminal possession," Mike said unceremoniously.

"Yup," Vince said.

"You know the paper came out today about the murder, saying she was being charged with it."

"Well, that's the paper. What counts are the papers in front of you."

Mike finished his review and said, "Where's the GSR report?"

"Don't have it yet," Vince said.

"So, fingerprints?"

"Not back from IAFIS yet."

"So, in other words, nothing that definitively puts the gun in her hand?"

"Not definitively, no."

Ken Ludlow came into the courtroom and nearly sprinted straight to Gary Becker, two pieces of paper fluttering in his hand as he did. Gary turned around.

"You got it?" Gary asked.

"Yup. Here you go," Ken said, some satisfaction in his voice.

Becker looked it over, smiled, and approached Mike and Vince.

"IAFIS," Gary said, "here's your copy."

At that moment, Deputy Wilson entered the back door, escorting Alina in handcuffs into the courtroom. Jason walked alongside her. Jason took a seat in the "peanut gallery," and Wilson walked Alina over to Mike and Vince. They moved so that they each sat on one side of Alina. Her handcuffs clunked against the mahogany table, and the sound startled her.

The court clerk entered the courtroom from the door behind the bench and took her seat at the small desk in front of the judge's bench. The court reporter followed, carrying her stenograph, set it down to the left of the bench, sat, checked the paper in the machine, turned and gave Vince and Mike a small wave.

The court officer entered next, rapped on the door and announced, "All rise! The Honorable Rodrick Vance, City Court judge presiding."

Judge Vance, all of the six-foot-five inches, slim and fit with a full head of jet black hair, entered in confident strides, his black robe flowing behind him, took his place at the bench and said. "Good afternoon, you may be seated." He then turned to Gary.

"Mr. Becker, are you ready to proceed?"

"Yes, Judge. The People are ready for the arraignment of Ms. Alina Weber."

He then turned to the defense table. "Mr. DeMarco, Mr. DeMarco, are you, one of you, representing Alina Weber?"

Mike answered. "We are, your honor."

"Well, I don't see a notice of appearance here," he said, flipping through the file.

"Right," Vince said. "This is a special appearance for the arraignment only. We haven't been retained at this point. We are prepared to proceed on that basis."

"Alright, I'll give you to five tomorrow to get that filed, or I'll appoint counsel."

"Fair enough, Judge," Vince replied. "Five tomorrow."

"Ok, let's move on," Vance said. "Mr. Becker?"

"Your honor, you'll find the prosecutor's information before you. We are, at this time, only charging the defendant with criminal possession of a firearm."

"And how do you allege did she possess this gun?"

"It was found hidden in her bed between the box spring and mattress when CSI investigated the scene of the murder. The murder of Dr. Weber. We also just received the IAFIS results confirming the defendant's fingerprints on the gun."

"Does the defendant have a legal right to possess the gun, pistol permit or any other legal basis to have it?"

"No Judge."

"Mr. DeMarco?"

"Not that we know at this time, Judge."

"Well, that, along with the affidavit of CSI Detective Cindy Maddison, is enough to move this along. Mr. DeMarco?"

Vince stood. Your honor, we enter a plea of not guilty. We waive a reading of my client's rights, we waive a reading of the charges, and wish to be heard on bail."

"Very good. Thank you, Mr. DeMarco. Go ahead."

"Your honor, Ms. Weber has no prior criminal history, and to date, there has been no direct link to her and the gun other than it was found in the marital residence. I have just been handed the IAFIS results, but certainly have had no opportunity to review them or have a defense expert verify them. The government may say a murder charge is coming, but until it does, this is a low-level felony where release on her recognizance or under probation supervision is proper."

"Alright, thank you, Mr. DeMarco. Mr. Becker?"

"Judge, I'm sure the Court is aware that this is but just the beginning of the investigation and prosecution for the murder of Dr. Raymond Weber. Further, it's the People's position that the gun as identified in Ms. Maddison's affidavit is in fact the murder weapon. Further, the defendant was the only person at the premises when the deputies first arrived at the murder scene. The defendant's entire family is in Russia, and she has no connections to this area whatsoever. She is unemployed and has a valid passport. All in all, the People submit to you, Judge, that not only is the defendant a flight risk but is a danger to society as well.

We ask the court to remand the defendant to jail pending trial without bail or, in the alternative, impose one million-dollars bail or two-million dollars bond. Thank you, Judge."

"She has a current passport, Mr. DeMarco?" The judge asked.

Michael turned to Alina.

"My husband has it. My husband took it. I don't know where he put it," she whispered to Michael.

"Your honor, my client advises she does not have possession of a passport, nor does she know where it is."

"That's not what I asked, Mr. DeMarco. Does she have a valid passport?" The judge said.

Alina nodded yes.

"Alright, alright. In consideration of the information before the Court and the full knowledge that this gun may very well be the murder weapon used in the killing of Raymond Weber, I hereby set bail at $500,000 or bond at one million dollars. Oh. And Jason, thank you for being here. I'm appointing you as her court advocate. OK?"

"That's fine, Judge," Jason said.

"Very well. Court is adjourned."

Vance struck the gavel, stood, all in the courtroom stood. He turned to leave.

"Excuse me, Judge!" Came a voice from the very back of the courtroom.

Vance stopped in his tracks. Everyone turned to see the short, stocky man with brilliant white hair in a dated three-piece brown suit that none of them had noticed until now. He was sitting directly behind Jason.

"Sir?" The judge said, more curious than patient.

"Thank you, Judge," the man said, standing. "Where do I pay the bail?"

"What?"

"Thank you, Judge. Where do I pay the bail?"

"Can you state your name, please?"

"Yes. I am Nikolai Volkov. Where do I pay the bail?"

"Mr. Volkov, did you hear the bail is $500,000?"

"Yes, Judge. Where do I pay it?"

Gary stared in disbelief; Vince exchanged glances with Mike and Alina.

"You pay at the Tompkins County Jail at the Sheriff's department."

"Cash is OK?" Volkov asked, deadpan.

Judge Vance was out of patience. "Sure, cash, gold, silver, whatever you want," he said, turning and leaving the courtroom shaking his head.

Vince turned to Alina; she just shrugged.

"Mike, go!" Vince said.

Mike bolted to the back of the courtroom to catch Volkov as he was leaving the courtroom with Jason and caught them just outside the door in the vestibule by the elevators.

"Hold on! Hold on!" Mike said.

Volkov froze.

"Mr. Volkov, I'm glad you're saying you'll post bail for my client, but I need to talk to you," Mike said.

"No time. No time to talk," Volkov said.

The elevator doors opened, and he stepped in backwards, crossed his arms tight in front of him. The doors closed, and he was gone. Mike turned to Jason just as Vince came out of the courtroom.

"Jason, who is that?" Mike asked.

"You heard him, Nikolai Volkov," Jason said.

"OK, sure, that's his name, but who is Nikolai Volkov?" Mike insisted.

"No idea," Jason said.

Vince listened, slightly amused. He knew Jason had strict confidentiality requirements as Alina's court-appointed therapist advocate and wouldn't say even if he knew.

As they stood there, Detective Ken Ludlow emerged from the courtroom with Alina, again in handcuffs.

"I'm taking her back up to the jail. It'll be interesting to see if that guy is going to pay the bail, or if he's just some kind of psycho."

"We'll follow you up," Vince said. "We want to know too."

Alina and Jason exchanged glances, and for the first time, she gave Jason a closed-lip half smile. He returned the same.

"OK, let's go," Ken said, breaking their glance.

"Mike and I will meet you up there," Vince said. "Jason too."

"No, it's OK," Alina said calmly and confidently.

Ken let out a chuckle, shaking his head. "Wow, she's pretty confident that guy's gonna pay her bail."

"I guess," Mike said.

"Come on, let's go Ms. Weber," Ken said, taking her by the elbow.

They left down the hall and exited the building through the back stairs to the sheriff's patrol cruiser. He put her in the back seat cuffed, and headed up the hill back to the jail.

"Well, I'm going up to the jail. I've got to arrange services for Alina," Jason said. "She's going to need pretty much everything, including a place to stay when she gets out."

Vince nodded. "Yeah. I think we should go up too, Michael. I just can't be so sure this guy is going to show up with half a million in cash to bail her out."

# CHAPTER TWELVE

**Two Years Prior**
**Annapolis**

Virginia couldn't have cared less about the night dinner cruise that had been arranged "just for her." She drove to the marina with Victor with the same enthusiasm and interest as if they were going to drop off laundry. She'd completely given up questioning Vic about the bills that were still pouring in, or anything about the boat whatsoever. To her, it was as much a lost cause as her marriage was at this point. She'd go along, playing the dutiful wife, putting on a smile, and just get through the night.

Bill and Rick sat on the back deck of Bill's Chris-Craft, finished their Heinekens, and walked off the boat and onto the dock walking toward Victor's Matthews at the end of the pier.

"Wow, Rick!" Bill said, "you got her all lit up! You get the generator to run?"

"How the fuck many times do I have to tell you, Bill; it's dead; it's fucking dead. I can't even get parts for that thing anywhere. I'm running the lights off shore power. We'll lose the lights as soon as we unplug from the dock."

"OK good. Well, at least it'll be a nice welcome for the Mrs. until we pull off. We all stocked?" Bill asked.

"Just like you said. Full bar, clams, shrimp, lobster," Rick confirmed.

"Who's coming?"

"All of us. All twenty-four plus his wife," Rick said. "The 'guest of honor.'"

"Good. Just like last time. How far out you think we need to go?"

"Just outside of view of any land. We'll head east toward Kent Island, get away from the Bay Bridge, then south toward Romancoke to more open waters."

"Perfect," Bill said. "Perfect. You start serving from the bar as soon as we leave the dock. Who's captaining?"

"Your favorite: Randy Spaulding."

"Good. He's never let us down," Bill said with satisfaction.

"Yeah. And he's going to "teach" Victor how to pilot the boat."

"Dark night already," Bill remarked, "no moon. Radar working?"

"It's as old as the hills, but it's working. Took some tweaking, but I got it going just fine."

"The engines?"

"You kidding? They're like brand-new," Rick said with a smirk, taking pride in his work.

When they got to the Matthews, John Holdings, the owner of the 1930 Frank Anderson yacht and the Blue Crab Shack, was already on board, with the buffet already set up, down in the salon; the full-serve bar ready on the back deck. John Miller, who owned the 1929 Elco and Tom Reynolds who had the 1950 Trumpy yacht, were also already on board the Matthews helping set everything up. It was already eight o'clock, and being early October, the sun had already set.

Victor and Virginia appeared from the parking lot, holding hands, or more accurately, Victor was holding her limp hand. Vic's heart was full of hope, with despair just waiting on the sideline. He knew tonight would be a make-or-break affair with his wife. She'd been very clear; it was her or that "goddamned boat." He was counting on tonight to fix things with his wife. So were the rest of the captains in the marina.

Bill and Rick greeted them at the dock at the Matthews, and they boarded together, Bill insisting Virginia board first. Victor went behind her, then Bill and Rick. Virginia boarded and looked around, hoping there would be some other women there for her to socialize with; there were none.

Captain Randy Spaulding was standing at the helm and greeted Vic and Virginia.

"You're going to captain this beauty tonight!" Spaulding said, tapping the top of the perfectly crafted thirty-six-inch dark mahogany spoked wheel. For the first time, Victor truly took in the magnificent beauty and complexity of the controls at the helm. In addition to the massive steering wheel, there were two large chrome levers for the transmissions of the engines and two of the same for the engines' throttles. The array and number of gages and electronics screens was dizzying to him.

Spaulding knew that look—the first time the owner of a boat this size and tonnage came to realize what he was really facing. He patted Vic on the back.

"Don't you worry. After tonight, it'll be child's play for you!" he said, then belly laughed.

This was no comfort for Victor. He turned and saw that Virginia was already at the make-shift bar at the aft of the deck with a cocktail to her lips. Little by little, the rest of the "committee" boarded the boat, each introducing themselves to Virginia and telling her what a great boat *AVENGER* was and that she sure was lucky to own her. As the last of the captains boarded the Matthews, Spaulding fired up the massive diesels, the boat gave a shudder, and the engines rumbled.

Rick hopped off the boat, disconnected the boarding stairs, untied the lines, and unplugged the power cord from the dock power box. The boat went pitch dark except for the red and green and little white navigation lights. A small battery-powered ceiling light lit the buffet in the salon.

Rick climbed back on board and yelled, "all clear!"

Spaulding shoved the shift levers full forward and inched the two throttle levers forward, and the fifty-ton vessel smoothly slid out of the slip. He spun the wheel hard to the starboard, and the boat gracefully turned to the right and out of the marina and into the Chesapeake. The water was dead calm and as dark as the moonless sky.

"Here," Spaulding said, "you take the wheel; she's your ship!"

Spaulding stepped off to the left, and Victor stood center of the wheel, placing his hands on the smooth polished mahogany.

"Now take it easy, make small corrections. She takes a minute to understand what you're telling her, so make a slight turn, wait until she responds, then another slight turn, until she's headed the way you want. Now, see that flashing lighthouse off to the port?"

"Where?"

"Off to the port, the left."

"Got it."

Spaulding slowly pushed the dual throttles forward bringing the boat to a fast cruise.

"Good. Head that way. What are you drinking?"

"What?" Vic asked, holding the wheel in sweating hands, What?"

"Drink! Drink! Whiskey, bourbon, vodka? What?"

"Ah. You got vodka and orange juice?"

Spaulding turned to John Holdings, "Captain here wants a screwdriver! And keep 'em coming!"

Victor turned to be sure Virginia saw him in command of the boat, but she was nowhere to be seen. He called out. "Virginia! Ginnie!"

"Hey hey hey!" Spaulding said, "you're going off course! Watch what you're doing!"

Victor thrust the wheel to the port, and the boat lurched hard to the left.

"I'll tell you what," Spaulding said as he nudged Victor away from the wheel. "You go enjoy yourself, and we'll do this some other time when there's not a party on board."

Victor, relieved, finished his drink in one long slug, went over to the bar, and made himself another one—a double. Then he went looking for his wife.

People were all over the boat. A few were out on the extensive bow, drinks in hand, several were down in the salon making up plates of food from the buffet. There he found Virginia, clearly already very intoxicated and rambling nonsensically to John Holdings, another cocktail in her hand.

Victor approached her.

"Ginnie, are you having a good time?" He asked, hopefully.

She ignored him, so he held her elbow," Ginnie, are you having--"

She jerked her arm away from him.

"I was, uh, I was just telling these guys how much, how much-"

"Hey Vic!" Spaulding called down the steps. "Get up here and captain your boat!"

"Yeah, go, Vic. Go captain your boat," Virginia said with slurring sarcasm. "I'm gonna get some fresh air."

Victor returned to the helm under Spaulding's supervision.

The wind had picked up; they were far from land, so the waves had swollen to three to five feet. Nothing the Matthews couldn't handle in stride, but she'd ride up on a swell, go over the crest, then the bow would come crashing down into the trough between the waves. For the seasoned boaters onboard, it was normal; for someone without "sea legs," it was a challenge to walk, especially drunk.

Virginia exited the side door from the salon onto the side deck. The deck along the side of the cabin was a little less than three feet wide, with a wooden railing on metal posts four feet high that ran the entire length of the yacht. Half-way to the bow was a gate that could be lifted to board the boat from the boarding stairs. It had been left open. Virginia found her way up the side deck toward the bow with one hand grasping the rail as the went, the other holding her drink. The boat was heaving in the waves. In the absolute darkness, she didn't see the open gate, and when her hand slid off the rail into the empty space left by the open gate, she tumbled sideways into the pitch-black water and screamed; her screams drowned out by the crashing waves, and thundering engines of the boat that vanished into the darkness. Even in her intoxicated condition, she managed to keep her head above water until the towering wake from the fifty-ton Matthews drove her twenty feet into the depths.

Victor heard and saw nothing of her fall but sensed something wrong.

He turned to Spaulding, "you take the helm for a while; I need to find Ginnie," Victor said.

"What do you mean find her?" he asked. "She's downstairs having a ball."

"Well, I just want to––"

Spaulding yelled down the steps to the salon, "Hey, down there! You got Ginnie?"

Holdings yelled back, "She's right here, and she's having a blast, aren't you, dear?" he said as if she was standing right beside him there in the salon.

Victor relaxed. Besides, she wasn't too happy to see him earlier anyway.

They cruised on.

Holdings turned to Rick. "She was trouble, that one. Real trouble."

Rick nodded in agreement. "Any stone crabs left?"

An hour, and twelve nautical miles later, Victor finally decided he had to talk to Virginia. He left the helm to Spaulding and went downstairs; it was crowded with members of the "committee" mulling around the buffet, some siting on the couch and chairs, and most fairly intoxicated.

"Have you seen Ginnie?" Victor asked Rick.

"She was here a minute ago. Didn't she come up on deck?"

"I didn't see her," Victor said.

"Well, she may have either gone to the bow or maybe she's passed out in one of the cabins. She was pretty wasted, you know."

A deep feeling of dread overcame Victor as he made his way to each cabin, looking for her. Nothing in the two forward cabins and nothing in the two aft cabins. His heart sank. He made his way back to the back deck and, from there, to the port side deck and, gripping the railing, made his way to the bow, which was now drenched in seawater. No one was on the bow. He then made his way to the starboard side deck and began carefully walking aft toward the back deck when his hand slipped off the railing, and he caught himself by grasping one of the metal stanchions. The gate was open, and he was now certain she'd gone overboard. He scrambled as fast as the slippery deck and lunging boat would allow and made it back to the helm. In his panic, he grabbed the shift levers and yanked them full-back, causing both engines to shift into reverse at high rpms. Spaulding immediately pulled the throttles back to idle and shifted the engines back into forward.

"WHAT THE FUCK ARE YOU DOING?" Spaulding yelled.

"SHE'S GONE! SHE'S GONE! SHE WENT OVERBOARD!" Victor shrieked.

"You sure?"

"YES! YES! YES! STOP THE BOAT! STOP THE BOAT!"

Spaulding pulled the engines into neutral and picked up the microphone on the VHF marine radio. "Mayday! Mayday! Mayday! *AVENGER* here. We've got a soul overboard."

The rest of the committee quickly began scanning the water with whatever flashlights they could find in a staged urgency.

Ginnie would never be found.

***

The Coast Guard search was called off a week later, and the "committee" organized a memorial at sea for Virginia. They let Vic choose which boat to use, and he decided that since it was the last place he saw her, the ceremony would be on his Matthews. Since they had no real idea exactly where she'd fallen overboard, they would just cruise the boat out onto the Chesapeake a few miles and shut down. They supplied sandwiches, drinks and the traditional wreath that Vic would toss into the sea.

They all boarded the Matthews, Vic at the helm with Spaulding and left the dock at 9:00 on Saturday morning. The sea was calm, with one-foot swells and a light breeze from the south.

"I can't tell you how much I appreciate this," Vic said.

"Hey, it's the least we can do. She was a special lady," Bill said.

"She sure was."

After a half-hour cruise, Spaulding tapped Vic on the shoulder. "You can shut her down now, Vic."

Vic stopped the engines, and the boat gently turned abeam the waves and began a gentle rocking motion. They all gathered on the bow of the boat, and Bill, the chairman, made a long, emotional speech and led the group in the Our Father. Rick handed Vic the wreath, and Vic tossed it into the water.

Vic was almost glad Virginia never had to know their house had gone into foreclosure. He'd made every effort to keep it from her, hoping he'd find away to save their home. But with the bills and debt that had stacked up, there was no way around it. Now, with his wife gone and no children, he decided to just deed the house over to the bank and be done with it. After all, he had the Matthews that was more than enough for him to live on by himself, and he knew that as far as he could tell, all the other boat owners in the marina lived on their boats year-round.

He renamed the boat *MY CHOICE.*

# CHAPTER THIRTEEN

**Present Time**
**Ithaca**

Nikolai Volkov had a driver waiting for him at the curb at the exit of the courthouse in a black Mercedes Maybach.

"Five hundred thousand," Volkov said as he got into the car. "Let's go."

They sped off to the jail.

When they got there, they glanced around the parking lot, and seeing no one, Volkov opened the case and counted out fifty sealed packs of one-hundred-dollar bills, one hundred bills in each pack—a half-million dollars. He closed the case, got out of the car, rushed in and paid the bail to the absolute astonishment of civil deputy Susan Alling. He used his alias, "Volkov," and, of course, gave the deputy a fake address. The deputy scribbled a receipt for the cash and handed it to him. He dashed out the door to his waiting driver, got in the car, and they sped off.

***

Ken arrived with Alina in cuffs fifteen minutes later. Volkov had already paid the bail and disappeared. As Ken entered the back door to the building, Deputy Wilson greeted him, and they walked together back to Wilson's desk in the "bull pen."

"Just un-cuff her and let her go; bail's already been paid," Wilson said.

"No shit," Ken said, shaking his head as he pulled the keys from his belt and began unlocking the handcuffs from Alina's wrists. "Unbelievable."

The cuffs came off, and Alina rubbed her wrists. "Thank you," she said meekly.

"Alright. You're good to go," Ken said, dropping the handcuffs on Wilson's desk.

"Thank you," she said again, now wondering where she would go.

"You have any stuff up here?" Ken asked.

"No. Nothing."

"OK then, you can go out to the lobby and call somebody to come pick you up."

"There is no one," she said, looking down to the floor. "No one."

*No one? No one, really*? Ken thought. There must be someone since that bail got paid.

"Well, you just go out to the lobby and have a seat. I have a feeling Jason, your court-appointed social worker, will be up in a minute."

She smiled, a tight-lip smile. "Ah, Jason. Yes."

"Take her out to the lobby," Ken said to Deputy Wilson.

"This way," Wilson said, gesturing down the hallway.

"And get her a candy bar or something from the machine out there."

Wilson ushered Alina down the hall, through the steel security door, into the lobby, had her sit in one of the ten black plastic chairs, went over to the vending machine, turned to her and asked, "Chips or candy bar?"

"Candy bar, please."

Just as the Snickers bar tumbled through the machine, Jason, Vince and Michael entered the front door to the lobby.

"There ya go," Wilson said to the group. "She's all yours...for now. Bail's paid."

"Nikolai Volkov?" Vince asked.

"What?"

"Was it Nikolai Volkov that paid the bail?" Vince asked.

"Don't know any Nikolai Volkov. He had the bail, and that's all we cared about," Wilson said. "Good luck with this one, Vince."

Wilson turned, opened the solid steel door and went back to his office.

"'Bail's paid,'" Vince said, mimicking Wilson. He turned to Alina. "Well, congratulations for getting bailed out," he said flatly.

"I don't understand," she responded. "I don't understand any of this."

"Your friend, Nikolai Volkov or somebody put up a half-million-dollars cash bail to get you out of jail until trial."

"Trial? Is there going to be a trial?" she asked.

Mike piped up. "Too soon to be talking about that," he said, giving Vince a disapproving side-glance.

"Right," Vince said, although knowing it was never too soon to be talking about a trial in a murder case.

"All we have is the criminal possession of a firearm charge," Mike reminded Vince.

Vince knew better. Vince was well aware of how District Attorney Gary Becker thinks and works after decades of battling cases with him in the courtroom. A murder charge was just around the corner. And the judge knew it too, or he would never have set bail so high. Vince had no doubt Gary was already assembling the Grand Jury to secure an indictment against Alina for murder in the second degree.

"Well, unless you need to meet with her now, I'm going to take Alina to get her settled in her room at the advocacy center and get some food," Jason said, tossing Alina a kind smile.

"That's fine," Mike said. "We'll need to meet with her tomorrow. She's had enough to deal with today."

"I agree," Vince said. "We'll see you tomorrow, Alina, and thanks, Jason."

"You bet," Jason said. You let me know when you want to meet with her, and I'll make sure she has transportation."

***

Meanwhile, Gary was in his office on the third floor of the courthouse in a rage. He called Ken Ludlow.

"What the fuck, Ken! What the fuck!"

"Whoa, whoa, calm down, Gary! Calm down!"

"The FUCK I'll calm down! A half-million dollars cash? Are you fucking kidding me?"

"I know, I know."

"And who the fuck is this Nikolai Volkov, shows up from nowhere?"

"No idea.

"Did you talk to him?"

"No. He came in, went to the civil window, paid the cash, got his receipt and took off fast. I can tell you that from the security tape, he was driving a quarter-million-dollar Mercedes."

"You get the plate number?"

"No. Didn't have a reason to." Ken said.

"I don't know who this Alina Weber is other than Ray Weber's now-widow, but it worries the fuck out me that she has some sort of connection where she can whip up a half-million dollars cash with the snap of her fingers. We're missing something here, Ken, something very serious."

"I can do more digging, Ken said. I can double-check the security tape to see if we caught the Mercedes plate number."

"Well, you do what you want, but I'm thinking maybe I'll have a talk with Tom Foster."

"The FBI?"

"Yeah. The FBI. That kind of cash screams money laundering. This might be above your pay grade, Ken."

"Ha! My pay grade? What I get paid? Everything's above my pay grade."

"I want you focused on the murder charge."

"Of course," Ken said, "Of course."

***

Raymond Weber met Helen Stinson in their last year at the State University of New York at Albany. He was pre-med; she was in the nursing program. Helen was intelligent, unpretentious, honest, and hardworking. When she met Weber, she was a full-time student and worked in the kitchen at one of the university's massive cafeterias to support herself. The attraction was fast and mutual. When Weber was accepted to medical school, they married and moved into a small apartment a few blocks from the school. Helen gave up her goal of becoming a registered nurse and then a nurse practitioner and went to work full-time to support herself and Weber though medical school. At nights, she often typed Weber's reports and helped him with his studies. The marriage went well throughout Weber's med school and internship, and with the promise of a lucrative future, they contacted a financial planner. Other than setting aside a certain amount of money each month into savings, they purchased life insurance on each other's lives: one-million on Helen's life and five million on Raymond's life. At their young ages, the premiums were low. They named each other as the sole beneficiaries of those life policies.

Toward the end of Weber's second year of residency, things changed. His arrogance was only overshadowed by his hair-trigger temper. He stayed out late at night and started drinking heavily. That led to arguments, more often than not leaving Helen bruised and battered. They divorced soon after he received his certification in Psychiatry.

When she learned of Weber's murder, she was flooded with emotions: elated that he was dead, shocked—but not surprised—that he was murdered, and thrilled she'd kept up the payments on the five-million-dollar life insurance policy on his life all these years. When she read his new wife was the prime suspect in the murder, she completely understood, and her heart broke for Alina. "Good for you, girl," she said aloud to no one.

She went to her study and found the policy in a metal lockbox in her desk drawer. She unfolded it, placed it on her desk, and smoothed it over with her hand. Then she opened her laptop, pulled up the insurance company website, and tapped the "file your claim" window. A blank form appeared on the screen, and she promptly began filling it out. When she was done, she accepted the "conditions" and hit "submit." The screen immediately opened a red box stating: "upload death certificate."

"Hmm," she said.

# CHAPTER FOURTEEN

When Vince got home, Mary Beth was sitting on the couch in her pink silk pajamas and bare feet, her legs tucked under her and one arm over the back of thc couch. On the coffee table before her was a bottle of Chateau Lafite Rothschild red wine, open with the cork in, and two wine glasses.

"My God, you're a sight for sore eyes," Vince said as he entered the room.

He took off his jacket, slung it over the chair facing the couch, and let himself practically fall into her arms.

"I missed you today," she said.

"I missed you, sweetheart."

They sat in silence, in each other's arms, Vince taking in her softness, warmth and delicate scent. She stroked his head.

"You're tired, aren't you, Vince." More a statement than a question.

"I'm OK. Quite a day."

"I can imagine," she said. "Can I ask you a question, and can you promise to be straight with me?"

Vince sat up. "Why would you say that? You know I'm always straight with you."

"Well, I'm sorry, I shouldn't have put it that way."

"Ok, ask me anything."

"Please tell me you're not going to take the Weber murder case."

"That wasn't a question," Vince said.

"Always the lawyer, huh, Vince? OK then, *are* you going to take the Weber murder case?"

Vince took a deep breath. "I… I…"

"Vince!"

"I don't know, Mary Beth."

"Well, then, let's talk about it. You know that defending a murder case is all-consuming, yes?"

"If it's done right, yes."

"That's the only way you know how Vince."

"OK."

"And we just bought a boat, did we not?"

"God, now you sound like me on cross-examination."

"Well?"

"Yeah. We did."

"And we did because?"

"I get it, sweetheart. I get it."

"I would never tell you what to do, Vince, but I can ask you to please think about giving this one to Michael. You know he can handle it."

"No question there."

"Have you spoken to him about it?"

"Not directly, no."

"Would you please?"

"Of course I will."

"Good. I love you."

"God, I love you too."

Vince picked up the bottle of wine and poured them each a glass.

"I'll call him tonight," Vince said.

"No no no. Tonight, you're all mine, all mine," she said, leaning in for a kiss.

***

Vince strolled into the office the next morning at nine-thirty.

The DeMarco law office was exactly what one would expect a law firm to look like hunter-green walls, dark mahogany wainscoting, floor-to-ceiling custom bookshelves packed with law books, and polished brass letters that spelled "DeMarco Law Firm, LLP" on the façade of the second-floor library above the main office.

Miranda was the one and only paralegal in the firm. She was experienced, dedicated, salty, had little sense of humor, and efficient.

"Well, good morning," Miranda said without looking up from her computer screen. "What are you doing here?"

He ignored the question.

"Good morning, Miranda. Mike with a client?" Vince asked.

"Not yet, but he's got Alina Weber coming in at ten."

"Really?" Vince said, surprised.

"Yes, really, Vince."

"OK, thanks."

"Sure."

Vince walked down the hallway to Mike's office door and knocked.

"Come in," Mike said.

Vince entered and plunked himself down in one of the maroon leather wingback client chairs facing Mike's desk.

"So you have a client coming in at ten, Miranda tells me," Vince said.

"Alina Weber. Jason's bringing her in."

"And you didn't tell me because?"

Mike paused and leaned back in his chair. "Because, Dad, you're supposed to be cutting back!"

"Yeah, but..."

"But nothing, Dad! I got this one. It's probably going to plea out anyway."

"Nope."

"Nope?"

"Yeah. Nope."

"Nope, what?"

"This isn't going to plea out. This is going to trial," Vince said convincingly.

There was a knock on Mike's office door; it was Miranda.

"Come in," Mike said.

"Uh. So, some guy just dropped this off, and I figured you'd want to see it right away," Miranda said.

She handed Vince a brown Gucci messenger bag.

"What's in it?" Vince asked.

"I didn't open it!" Miranda said.

"Of course not," Vince said as he gingerly opened the valise. He pulled out a plain brown envelope and a small black leather satchel. He set the satchel on Mike's desk, opened the envelope first, and slid out its contents.

"What is it, Dad?" Mike asked.

Vince was incredulous. "It's one of our retainer agreements."

He flipped through the pages.

"It's signed by Nikolai Volkov, and where the type of case blank is, it's typed, 'for the defense of Alina Weber' and..." he paused, staring at the third page.

"What, Dad?"

Miranda was trying to see over Vince's shoulder.

Vince looked up at Michael.

"The retainer amount entered is $100,000.00."

They all looked at the satchel.

"Open it!" Miranda said.

Vince picked up the satchel, opened it, and pulled out ten packets of ten one-thousand dollar bills. He set them down on Mike's desk and ran his hand all through the satchel and the message bag, looking for anything that might give him more information. There was nothing.

"I think I want a raise," Miranda said, turning on her heel and leaving the room.

"How did he get our retainer agreement?" Vince asked Mike.

"You kidding? It's on our website," Mike said.

"Ah."

"But Dad, have we even agreed to take this case? I mean, we usually don't take any case, especially a case this serious, without an initial consultation with the client and some background search."

"It appears we've been *selected* to take the case, doesn't it," Vince said.

"Well, Nikolai Volkov signed the retainer. The money is here, but we haven't signed the retainer agreement yet, so until we do, we're not retained. I don't care who 'selected' us."

"I couldn't agree more, Michael. So, she's coming in this morning, yes?"

"Yup."

Michael's intercom pinged. "She's here," Miranda announced.

"Bring her back," Mike said.

"Jason's here too. You want him?"

"Not yet," Mike said. "Maybe later."

Miranda offered Alina and Jason coffee or water; both declined, and Miranda ushered Alina into Mike's office. Jason sat on the couch in the waiting area. When Alina entered, both lawyers stood and greeted her.

"Have a seat, Alina," Vince directed.

She sat and nervously adjusted herself in the chair, then settled with her hands folded in her lap, looked at Vince, then across the desk and fixed her gaze on Mike.

"How are you holding up, Alina?" he asked.

"I'm OK," she said.

"Alina, we haven't decided whether or not we're going to take your case yet," Vince said, getting right to the point. "We have a lot of questions we need answered."

"I understand."

"So, today, we received this," Vince said, handing her the retainer agreement signed by Nikolai Volkov.

"Who is he?"

Alina stalled, slowly looking at each page, pretending she was reading.

"Alina?" Mike said, prompting her.

"My father. Maybe my father."

"Maybe your *father*?" Mike said, astonished.

"Well, that certainly answers one question!" Vince said.

"Yes. But I've never seen him until now. Only photos."

"Why is that?" Michael asked.

"Because my mother has always told me he died when I was a baby. Photos. I had photos of him."

Vince and Mike exchanged glances, Vince raised his eyebrows.

"Are you're sure he's your father?" Vince asked.

"No," she said, "I'm not."

"Well, here's an obvious next question," Vince said. "What's your maiden name, Alina?"

"Sokolov, Alina Inessa Sokolov."

"This is getting nowhere," Mike said. "Let's get to the merits of the case."

Vince nodded.

"Alina, before you tell us anything, you need to know that anything you tell my father or me is in strict confidence," Mike said. "Even though we haven't formally accepted your case, you're covered by the attorney-client privilege as we sit here, the strongest privilege known. We will never disclose anything you tell us unless it's to advance your case, to help you and unless we discuss it with you first and you agree. Do you understand?"

"I think so. It's all a secret."

"Well, kind of," Vince said.

"I didn't kill my husband," she said matter-of-factly.

"Alright," Mike said. "I want you to try to relax, take a deep breath, and try to tell us what happened that day."

Alina closed her eyes, began wringing her hands, and little by little began taking deep breaths. Her head dropped, and tears flowed from the corners of her eyes. Then she began to tremble.

"Alina, are you alright?" Mike asked.

Vince looked Mike in the eyes and shook his head no.

"Alina?"

Her trembling worsened, and Vince worried she might convulse.

"OK, Alina. OK. We don't need to talk now; it's OK."

Mike pressed his intercom button.

"Miranda, bring Jason back here and a water for Alina, quick!"

Jason rushed to Alina while Miranda grabbed a bottled water from the kitchenette and ran to Mike's office. Jason crouched in front of Alina and took both her hands into his.

"Alina."

She opened her eyes and, seeing Jason, began to sob uncontrollably. Jason resisted the urge to hold her but kept her hands in his and turned to Vince.

"We need to end this meeting," Jason said. "I'll take her to my office and get her calmed down, take her to get something to eat and then we'll see how to proceed from there."

"We understand," Mike said.

Vince nodded.

Jason pulled a tissue out of the tissue box on Mike's desk and handed it to her.

"She's got to be traumatized, probably more than we know," Jason said. "No matter what, she was there when all this happened." He turned to Alina. "Come on, Alina. Come with me, we're going to talk, OK?"

She nodded, not a word spoken, got up from the chair, and turned to leave the office.

"We'll stay in touch," Jason said.

"As soon as possible," Mike said.

Miranda walked in with a bottle of water in her hand. Mike waved her off.

They waited until they were sure Alina and Jason were out of the office.

"We actually know less now than we did before she came in," Mike said.

"Well, you know, as lawyers, we get numb to the emotions people suffer. We look at it as facts, and the law and how the facts fit the law."

"Sounds cold as hell, dad."

"Well, it is. And we have to be that way not to get sucked into the emotional side of things, whether it's a divorce case, domestic violence case or..."

"Murder," Mike said, punctuating the sentence.

"We weren't going to get any information from her in her state today."

"Unless it was an act," Mike said.

"Now, who's cold as hell?" Vince said.

"Ha. Well, what do you think, dad? You think she killed Weber?"

"Doesn't really matter what I think, but for the sake of conversation, it looks pretty bad, doesn't it?"

Mike leaned back in his chair and crossed his arms in front of himself. "Yup, looks pretty bad for her. I mean, she was alone with Weber when it happened; she hid the gun under her mattress, fingerprints all over the gun. Isn't that about all Gary has to prove at trial?"

"He has to place the gun in her hand. Directly in her hand when the shots were fired," Vince said.

"That's easy enough: blow-back, GSR and blood spatter."

"That would pretty much clinch it, don't you think?"

"Self-defense?" Mike proposed.

"Hm."

"Maybe 'battered spouse syndrome.'"

"Now, that might be," Vince said. "We need to talk to Jason about that. He'd be the one to help establish it."

"I don't think anybody that knew Weber didn't know what a temper he had. We need to talk to Alina about it."

"We do," Vince said.

# CHAPTER FIFTEEN

Jason's phone sounded just as he was bringing Alina into his office.

"Jason, it's Mike."

"Yeah, Mike, what can I do for you?"

"How's Alina doing?"

"She calmed down quite a bit in the car. Stopped at McDonald's and got her something to eat."

"OK, so listen, I'd like you to think about the battered wife syndrome in your therapy with her."

"Mike, it was already on the top of my list," Jason said.

"I should've known, Jay. Thank you. Let me know when she's up to talking with us."

"Will do."

Jason clicked off the call, had Alina sit in the single wide polyester barrel chair facing his vintage maple desk, and took his place behind the desk. The office was fairly bare, undecorated and intentionally so. Jason didn't want anything in his office to distract a client from clear thought.

"Are you feeling better?" Jason asked Alina.

"Yes, thank you, Jason."

Jason got up, went over to the coffee stand to the left of his desk, poured two cups of coffee.

"Sugar?"

"Lots," she said.

He gave her the coffee, sat back at his desk, took a sip and waited for her to speak.

Nothing came.

"So, can you tell me a little bit about your relationship with your husband?" Jason started.

"What do you want to know?"

"Well, did you get along well? Did you argue?"

Alina took a long pause to think.

"Sometimes good, sometimes not so good."

"OK, tell me when it was good."

Another pause. She couldn't understand why she was having so much trouble remembering anything.

"I had a good home, plenty of food, clothes."

"Of course. I understand. But tell me about your relationship with Dr. Weber. Was he kind to you? Did he become angry with you?"

She thought hard.

"He worked very hard," she said.

"OK, doctors work hard," Jason confirmed.

"And he would come home very tired."

"I understand."

"Sometimes, we would have a nice dinner together and have a good night."

"And other times?" Jason probed.

She took a deep breath, and Jason noticed she was wringing her hands.

"Not so good," she said.

"OK, so not so good. What does that mean?"

She tightened her lips as if she wanted to stop herself from saying anything.

"Sometimes very very bad, Jason."

"Tell me."

"He would drink. When he drank, things were bad."

"He struck you?" Jason speculated.

She nodded her head and looked down into her lap.

"Often?"

She nodded again.

"Did you tell anyone?" he asked.

"Oh no! I never told anyone."

"You're family? Did you tell anyone in your family?"

"Oh no. never."

"Friends?" Jason asked.

"I have no friends. He told me he was the only friend I needed."

"Any friends on FaceBook or anything like that?"

"No, I wasn't allowed. He told me that such things were nothing but trouble."

"So you told no one?"

"Ha. Even if I told my mother, she wouldn't believe me. I told no one."

Jason surreptitiously slid a book on his desk to the edge and let it fall.

Alina jolted, gasped and almost stood from the chair, spilling coffee onto the floor.

"Are you OK?" Jason asked.

"I'm OK; I'm so sorry," she said, embarrassed, "I can clean it."

"Hm. No worries," Jason said. "It's just a few drops." *Hyper vigilant,* he confirmed to himself.

She tried to compose herself, still sitting on the very edge of the chair.

"This is all my fault," Alina blurted. "This is all because of me," she said, a tortured look on her face.

For the first time, Jason noticed how bloodshot her big hazel eyes were and how pale her face was.

"Why do you think that, Alina?"

"I don't know. I do stupid things. I say stupid things."

"What kind of things?"

"He would tell me, 'You're stupid!' 'You're an idiot!' 'You're a whore!' And see? I spilled the coffee!"

"I see. Why do you think he said these terrible things to you?"

"Because he is a doctor, and I'm nothing. And because he paid money for me to be his wife, like a prostitute."

"Are you a prostitute?" Jason asked.

"No. I don't think—"

"Of course, you're not, Alina. You were his wife."

She shut down.

"Alina?"

"I was his wife. I tried to be a good wife, but I did things that made him mad."

"What kind of things, Alina?"

Her mind went blank. She took a long sip of coffee, trying to think. Nothing came up.

"I don't know," she said. "I can't remember."

"Are you having difficulty remembering things, Alina?"

"I am."

"How do you sleep?" Jason asked.

"What do you mean?"

"Do you usually sleep well at night? I mean before all this happened."

"I sleep sometimes, but I have nightmares."

"Well, we all do, I guess," Jason said.

"I'm tired, Jason."

"I can tell. Are you ready to go back to your room at the center?"

"Yes, please."

Jason took her back to the center and drove back to his office to call Mike.

Vince and Mike were still in Mike's office discussing the case and a couple other cases Mike had taken over from Vince. Mike's intercom sounded.

"Jason on line one," Miranda announced.

Mike picked up the receiver and pushed the line-one button.

"Hey, Jason, how's she doing?" Mike asked.

"I'd say a little better. We had a good talk...a good start anyway."

"And?" Mike asked.

"Textbook battered wife syndrome," Jason said. "I can't say if she had anything to do with what happened with Weber; all I can say is that in just one session, she showed all the elements of the battered wife syndrome, right down to the letter."

"That's what we need to know, Jay. Thanks."

"You bet. I'll be working with her until the judge says otherwise," Jason said.

"All the elements right down to the letter, you said?"

"Yup. Unless she's a great actress," Jason said.

"Hm. A great actress. Let's hope not."

Mike hung up the phone. "You catch that?" he asked Vince.

"Oh, yeah. I caught it."

"So, maybe there's a battered wife syndrome defense," Mike suggested.

"Double-edged sword, you know."

"Yup. If we were to use that defense, we'd have to admit she killed him and use the syndrome as an absolute defense."

"That's worse than tossing the dice," Vince said.

"So, what do you think, Dad?"

"About what?"

"About taking the case."

"Ah. Right. Well, I have to say, it all intrigues me, and there is the retainer that's twice what we usually get. You?"

"I'd like to take it. If you want, you can second-chair me."

"Hm. That's quite the offer." Vince said, never having ever been a second-chair to anyone. But the thought of being second-chair to his son was thrilling for him, and he felt so proud knowing Mike was fully capable of a case this challenging. "If you take this case, I'd be honored to be your second-chair."

"Then we have a deal," Mike said, reaching his hand across the desk to Vince.

Vince shook his hand and said, "I'm proud of you, son."

"I know, Dad. I love you."

"Well, I better go break the news to Mary Beth."

"Good luck with that, Dad."

"I may need it, son."

Vince had some comfort in the fact that the only charge so far was the criminal possession of a firearm. That comfort evaporated when he thought about Gary and the way Gary prosecuted his cases—relentless, and there was no doubt a murder charge was imminent. Nonetheless, he had faith in his son. He'd be proud to be second chair to Michael.

# CHAPTER SIXTEEN

After he shot and killed Dr. Raymond Weber, Weber's attorney, Sean Adams avoided his small one-room plain office on the second floor of the Circles Building that had been converted into low-rent basic offices mostly occupied by young, inexperienced lawyers like him. He was in full panic mode, spending time going from bar to bar, trying to sleep for a few hours in his studio apartment, wandering around downtown, and trying to formulate a plan—any plan. So far, as far as he knew, there appeared to be no suspicions his way; that was the good news. The bad news was that it was a foregone conclusion, apparently, that the girl he was hopelessly in love with, the girl he had done this for, was the one and only suspect in the murder.

After two days of purgatory, he ended up at his office. On the door was a pink post-it note: "Mr Adams, I understand you were Dr. Weber's lawyer. I'm his ex-wife, and I have to talk to you. My phone number is 555-246-7344." It was signed, Helen Stinson (Weber). He stopped dead in his tracks, paralyzed, staring at the little note. Instead of a thousand questions popping up into his head, he went completely blank. After a few seconds, he pulled the note off, unlocked the heavy oak door, went into his office, sat behind his used school teacher's desk, and stared at the note. What could she possibly want? He wondered. He turned on his laptop, and it showed a hundred unread emails. He scrolled down the screen, not realizing he'd stopped breathing, in absolute terror of what he might see. To his relief, the majority were junk mail, with only about twenty from clients and other attorneys. None of them referenced Weber. He snapped it shut, got up, went over to the cheap Mr. Coffee he kept in the bathroom behind his office, and made a pot of coffee. Weber's ex obviously knew he was Weber's attorney, and she clearly knew where his office was. He realized he had no choice but to respond to her.

He rinsed his one-and-only coffee mug and poured himself coffee, then went back to his desk, picked up the receiver to his desk phone, held it to his ear, and paused. He felt his heart pounding in his chest and took a deep breath. He dialed the number. It rang four times, and with relief, he was about to hang up when he heard her answer.

"Hello," she said, with a sing-song, pleasant ring to her voice.

His chest tightened, and he took another deep breath to relieve some of the tension.

"Ms. Stinson?"

"Yes, this is Helen Stinson."

"This is attorney Sean Adams. I got your note. What can I do for you?"

"Oh! Thank you for calling. You're Dr. Weber's attorney, or at least you were his attorney, right?"

Sean took a second to consider what exactly his relationship with the dead man was at this point.

"I am…I was…I…"

"Good," she said. "When Raymond and I were married, we took out life insurance policies on each other's lives."

"OK, and?" he asked.

"Well, I still have the policy, and it's still in effect. So, I would like your help in filing the claim."

"It's still in effect?" he asked, doubtful.

"It is. I've been paying the premiums since the divorce. I figure you'd be the right person to help me file the claim, since you're his attorney. I imagine you'll be probating his will and all."

*Fuck,* Adams thought to himself. *His estate? How can I possibly? I mean, I can't possibly do that. Some other lawyer should probate his will.*

Then it dawned on him. No other lawyer should be involved in this; it might expose him, questions would be asked. He would have to take control.

"Yes, Ms. Stinson. I'll be happy to handle that for you. I'll need the original policy."

*Wow,* he thought, *not only is she rid of her ex, now she's getting his life insurance money.*

"What can you tell me about the policy?" he asked. "The company, the limit? Do you know what kind of policy it is?"

She was quick with the answers, too quick. "Presidential Life and Casualty, whole life."

"The limit? The amount of coverage on his life?"

"Five million," she said. "Five million."

He stifled a laugh, smiled, and thought, *now there's a motive for murder.* His mind immediately began thinking of ways to bring the suspicion of Weber's killing onto his ex-wife, Helen. *Classic,* he thought.

"I can bring it to you tomorrow," she said.

"Fine. Tomorrow at ten would be fine."

"I'll see you then," she said.

He hung up the phone, leaned back in his chair, a grin on his face, his mind churning.

***

Michael was buried in discovery documents in a labor dispute case he was defending when his intercom sounded.

"Mike, it's……hehehehehe. It's…"

"What is so funny, Miranda?" Mike asked, a touch of impatience in his voice.

"Sorry. It's Sean Adams on line one for you, of Franklin, Underwood, Camden, and Knowles, LLP hehehehehe."

"What is so funny?"

"I'm sorry, Mike. I just pulled up their website."

"Yeah?"

"Yeah, and their logo is FUC & K, LLP. Hehehee."

"Take a break, Miranda. I'll take the call," Mike said, shaking his head.

"Mr. DeMarco, this is Sean Adams. I was Dr. Weber's defense counsel in a malpractice case."

"Yes, sir. What can I do for you?" Mike asked.

"Well, you see, I understand your firm is defending his wife, Alina, against murder charges."

"No. There are no murder charges," Mike said flatly.

Adams was taken aback. "I'm sorry. No charges?"

"No murder charges. So far, just criminal possession of a firearm."

*My fucking gun!* Adams thought.

"Why?" Mike asked. "What's this about?"

"Well, I don't know if it means anything really, but I got a call from Weber's first wife, Helen Weber."

"And so?"

"It seems she's got a five-million-dollar life insurance policy on Weber's life they took out during their marriage, and it's still in force."

"OK."

"Well, I thought that if there ever were murder charges brought against Alina…or anyone that—"

"Wait. I don't understand," Mike said, "What's your point?"

"Motive."

"Motive?"

"Sure. Motive. A five-million-dollar life insurance policy on an ex-spouse. Wouldn't that be a motive for murder?"

"I see," Mike said. "Do you have any proof this policy really exists? I mean, really, maybe she's just nuts."

"Well, maybe, but I'll know soon enough. She's bringing me the policy to file the claim. And I guess I'll be probating Dr. Weber's estate, too."

Mike paused.

"Was there a contingent beneficiary named on the policy?"

"I don't know," Adams said. "But I'll know when I see the policy."

"I see. Well, I appreciate the call, Mr. Adams."

"Sean, please. Call me Sean."

"Sure. I appreciate the information. Please keep me posted. And call me Mike."

"Got it. Thank's Mike."

Mike clicked off the call and dialed Vince, who was in the car with Mary Beth, headed down to their newly acquired boat.

Vince tapped the incoming call on the screen in his car.

"Yeah, Mike."

"Hi, Dad. Where are you?"

"We're headed down to the boat."

"Sure. Trying to get away from it all, huh?"

"Trying, Apparently not succeeding."

"Well, sorry, Dad."

"There's an every-Friday-night meeting of boat owners at the marina, and Mary Beth and I are going to attend our first one tonight."

"Gee, sounds real fun," Mike said, slightly sarcastically.

"I think so," Vince said, turning and smiling at Mary Beth.

She rolled her eyes.

"I just got an interesting call from Sean Adams," Mike said.

"Ah. Wasn't that Weber's malpractice attorney?"

"Yeah. But listen to this. He got a call from Weber's ex-wife who says she's got a life policy on Weber's life for five million, and the policy's been in force since before they got divorced, and the ex is still the beneficiary."

"Interesting, but so what?"

"He thinks it shows that Weber's ex had a motive for the murder."

"Ah. So, let me get this straight. He thinks that because the ex has a five-million-dollar life policy on Weber's life and the ex is the beneficiary, it might clear Alina, right?"

"I guess so," Mike said.

"Well, then, I guess he doesn't know much about divorce law in New York," Vince said, shaking his head.

"Clearly, Dad. Clearly," Mike said. "Well, have a good time at the boat, and say hi to Mary Beth."

"She's right here," Vince said.

"Hi, Michael!" Mary Beth said.

"Hey Mary Beth, have fun."

"Oh, Sure," she said.

Vince clicked off the call.

"Are you OK, Vince?" Mary Beth asked. "You look upset."

"Well, despite the 'favor' this guy Adams thinks he'd doing for Alina, he just gave the DA five million reasons Alina would want Weber dead."

"I don't understand," Mary Beth said.

"Think about it," Vince said. "In New York, a divorce decree is an automatic revocation of a beneficiary designation to the ex-wife, right?"

"Right."

"So even if his ex-wife is still the beneficiary on the policy, she gets nothing from it."

"Exactly."

"Then what happens?" she asked.

"If there's a contingent beneficiary—a second person named—that person gets the money. If there's no contingent beneficiary named, it goes into the estate."

"Is that a problem?"

"Weber had no children, so it's likely there was no contingent beneficiary named on the policy."

"So, then what?"

"Alina would be the only natural heir."

"Oh God," Mary Beth said. "That means Alina would've gotten the five million."

"Yup. Five-millions motives."

"So this guy, Adams, thinks he's doing Alina a favor when he may have just assured her conviction."

"Could be. Just might be, Mary Beth," Vince said, taking a deep breath.

"If Gary's smart enough to figure this out," Mary Beth said with a half-smile.

"Well, I wish Adams had kept his mouth shut," Vince said, shaking his head.

"Yup," Mary Beth said.

# CHAPTER SEVENTEEN

District Attorney Gary Becker was sitting at his office desk with stacks of files on the desk to his left, to his right, and on the floor, focused on the latest to come in: a witness in a case he was to take to trial in two months had contacted him and provided proof in the way of text messages that the girlfriend of the defendant on trial had offered the witness five-thousand dollars to recant his prior statements. That person was now in custody.

Gary's intercom buzzed.

"Tom Foster on line two," the receptionist announced. "FBI Special Agent."

Gary picked up the phone and pushed the line two button.

"Hey, Tom."

"Hi, Gary."

"So what do you have for me?" Gary asked.

"Nothing final yet, just some preliminary stuff."

Gary had handed the murder weapon in the Weber case over to Foster to submit to the Bureau of Alcohol, Tobacco, Firearms and Explosives (ATF) National Tracing Center (NTC), the United States' only crime gun tracing facility. NTC's mission was to conduct firearms tracing to provide investigative leads for federal, state, local and foreign law enforcement agencies.

"We got a hit on the serial number; nothing had been done to it, no attempt to file it down, and it was clear as a bell. Gun was sold new out of Fort Lauderdale fifteen years ago. Legit gun dealer legit registered buyer. From there, it was sold three more times, all in the Fort Lauderdale area, then once in Miami. There, the legit trail ended."

"Legit trail?" Gary asked.

"Yup. Stolen from a retired school teacher in Ocala, turned up in a drug deal gone wrong in South Miami."

"So the police had it, right?"

"Yeah. They did…or should have."

"What do you mean, should have?" Gary asked, sitting up in his chair.

"Don't know yet."

"Don't know yet?" Gary asked.

"Nope. Don't know yet. Look, you know, seized guns can be destroyed, sold at auction to licensed dealers, or even handed over to the military."

"Sure, but there'd be records of any of those transfers."

"Yes, Gary, and that's where the trail stopped so far. We don't know what happened to it after that, but we're still checking."

"So, I need to know how that gun got into Alina Weber's hands; that's all I need to know. Certainly, she had no legitimate reason to have it, and Dr. Weber had no legit reason to have it in his house, so it had to be stolen one way or another."

"Or maybe, just maybe, some other person brought that gun there that night," Foster suggested.

Gary paused.

"What do you mean, Tom?"

"I don't know, I'm just talking out of turn. I'm just saying, is there no possibility that a third person, an intruder, was there when Weber was shot? That maybe it was a robbery gone wrong?

I mean, seriously, Gary, what are the odds that a doctor or his wife would have a stolen gun in their home? Really, Gary."

"You've seen the reports, Tom."

"Yeah, I've seen them. Sometimes, when something seems so easy on its face, it really isn't. You know that."

"Well, you got me thinking, but I'm ready to indict."

"That's your call, Gary, but I'm still waiting on the NIBIN report."

"NIBIN," Gary said. "How long on that?"

"Depends."

"On what?" Gary asked.

"If that gun's ever been used in any crime before the Weber murder. If it has, we'll have digital images of the spent bullets and maybe cartridge cases that were found at crime scenes or test-fired if it was a confiscated weapon."

"What about that drug deal gone wrong in South Miami?" Gary asked.

"That could very well tie this gun to that shooting, but where would that get us?"

"It might get us to where I want to get, to trace the gun backwards to where Alina Weber got it."

"Maybe," Foster said.

"Well, you've got the spent bullets and the casings we found at the scene."

"Yup. They've been sent to NIBIN, and their turn-around is usually about seventy- two hours. Haven't seen them yet, though."

"OK. Keep me posted, Tom."

"Always, Gary."

***

"They're here again," Rick said. "I don't like it."

"Yeah. Here they come," Bill said as he and Rick sat on the back deck of Bill's boat, sipping their Budweisers. "What they got with them?"

"Looks like cleaning stuff."

"Yeah. Does. Shit."

Just then, Vince and Mary Beth, both dressed in blue jeans and matching gray sweatshirts and boat shoes, walked onto the pier behind Bill's boat. They smiled and waved at the men. Bill and Rick put on smiles and waved back.

"Got some cleaning to do, I see," Bill said cheerily.

"Sure do," Vince responded.

Mary Beth just looked down the pier.

"You need anything, you just let us know," Rick said.

"Oh, I think we've got everything we need for now."

"Alright then. Have a good one," Rick said.

"See you at the meeting tonight," Vince said.

The smiles disappeared from Bill's and Rick's faces. They waited until Vince and Mary Beth were far down the pier.

Bill turned to Rick with a stern look, "You sure you got everything?"

"Yeah. I'm sure. I'm pretty sure."

"Pretty sure? You better be damn sure, Rick. They're going down there to clean the boat for chrissakes."

Rick took a deep breath. "I'm pretty sure."

"Fuck," Bill said. "And I bet she'll be at the meeting tonight. I don't like it. I don't like any of it. I don't like it at all."

Captain Randy Spaulding approached Bill's boat.

"Permission to come aboard?" He asked perfunctorily as he walked up the boarding steps onto the back deck.

"Beer?" Bill asked.

"Sure," he said, taking a seat at the small round table next to Rick.

Bill went belowdecks and reappeared with an open bottle of Budweiser and handed it to Spaulding.

"You're not going to like what I have to tell you," Spaulding said, taking a slug of his beer.

"Yeah?" Rick said.

"Yeah."

"So?" Bill pushed.

"That girl, Mary Beth. Not good."

"Well, we know she's going to be trouble, Randy."

"Oh God, you have no idea," Spaulding said.

"Well, just fucking get it out, would you!" Rick spouted.

"You're not going to believe this. She's fucking FBI."

Bill spat beer across the table and started coughing uncontrollably.

"FBI?" Rick asked, "Are you fucking kidding me? Are you sure?"

"Positive."

"How?" Bill asked, finally clearing his throat.

"John," Spaulding said.

"John?" Bill asked.

"Yeah. John Holdings—our John," Spaulding said.

"I don't get it," Rick said.

"The Volkov thing,"

Bill and Rick exchanged glances.

"John recognized her from the thing up on the Potomac."

"Is John still doing the 'Volkov' thing?" Rick asked.

"Of course, he is," Bill said, leaning back in his chair and rubbing his forehead.

"How does he know?" Rick asked.

"You know I run everybody by Holdings, everybody that comes into this marina. When I mentioned her name, I thought he was gonna have a fucking heart attack."

"What'd he say about her?" Bill asked.

"Not much. Just that he'd had a run-in with her before, and it wasn't pleasant."

"Well, then, I guess we don't have much of a choice, do we, gentlemen?" Bill said.

"Never thought we did," Rick said. "This seals it."

Spaulding nodded.

***

Vince and Mary Beth sat in the white wicker chairs on the back deck of the Matthews with a bottle of Gravner Anfora Pinot Grigio on the round wicker table in front of them. Vince refilled Mary Beth's glass.

"Shall we go?" Vince asked Mary Beth.

"So we shall," She said. "What time is the meeting?"

Vince looked at his watch, "in about an hour."

"Good. Just enough time to enjoy this," she said, taking a sip from her newly filled glass of wine.

At six o'clock that Friday evening, John Holdings left his 1930 sixty-foot Frank Anderson Yacht, left the marina and headed down to the Blue Crab Shack downtown. It was his job to provide the buffet dinner for the marina committee meetings in the large room in the back of Rick's shop every Friday. By eight, everything was ready: stone crabs, oysters, corn on the cob, Cole slaw, and a full bar. By nine, the full committee was in place, including Vince and Mary Beth. Chairman Bill Brown started the meeting.

"Welcome, folks! Glad you could all make it. We'll have a short meeting, then enjoy the great buffet John's set up for us."

"Where is he?" John Miller, the owner and captain of the 1929 75' Elco asked.

"Well, things came up, and he couldn't make it," Bill responded. "But let's have a hand for him for this great buffet."

They all applauded.

Bill went on. "And let's have a hand for Vince and Mary Beth, the proud new owners of *MY CHOICE*, that grand old Matthews at the end of Pier One."

They all turned to Vince and Mary Beth and again applauded.

"And!" Bill interrupted, "I'm happy to announce that John has offered to cater an evening dinner cruise on *MY CHOICE*, including an open bar to welcome Vince and Mary Beth to our marina if they're willing to host on their yacht."

Vince and Mary Beth were taken aback by the over-enthusiastic applause.

Vince waited for the applause to calm down and said, "Of course, that would be our pleasure!" He turned to Mary Beth, who sat stone-faced. "Right, honey?"

Mary Beth was having trouble counting all the red flags popping up in her head.

***

This would be the first night Vince and Mary Beth would spend on *MY CHOICE*. After a cursory cleaning of the cabins, they returned to the car and brought out clean sheets, blankets, pillows, a small cooler, and two grocery bags filled with drinks, sandwiches, and snacks. They made up the double bed in the main stateroom at the aft of the boat. Despite the cleaning they'd done and the freshly made-up bed, Mary Beth was still not convinced.

"Vince, maybe we should stay at a hotel tonight."

Vince chuckled. "Why would we do that? We're all cozy and warm here, and the bed's all made up."

"I just don't feel good about it, Vince."

"Why, honey?"

"I don't know, Vince. It's just a feeling."

Vince almost said something about female intuition, but he thought better of it.

"Is it anything specific?" He asked.

She sat on the bed; he sat beside her and put his arm around her.

She took time to think.

"Is it the diesel smell?" He asked.

"No no no. Once we aired everything out and got the heat to work, it's not that."

"Then what?"

"Wait!" Vince said, suddenly popping up off the bed. "Wait!"

He bolted out of the cabin and returned minutes later with an open bottle of Chardonnay and two wine glasses. "I forgot we brought this!"

He handed her a glass and poured, then filled his own glass.

"To our first night aboard!" He said, clinking his glass to hers.

"I surrender," she said, shaking her head. "To our first night on board."

Mary Beth now knew she had work to do, but this was not the time to drag Vince into it.

After a quick breakfast at Captain Moe's diner downtown, the next morning, they returned to the boat with groceries for lunch and dinner and got back to cleaning. Sunday morning, they took their first cruise out onto the Chesapeake. The sky was clear blue, the winds were moderate from the northwest, and the seas were a choppy two to three feet. Vince was pleased with its performance, as the hefty fifty-ton hull hardly felt the three-foot swells. As the wind picked up around noon and the waves built up to four feet, Mary Beth called it a day, and they went back to the dock and made lunch together.

Rick and Bill, sitting on Bill's back deck, watched as Vince smoothly backed the yacht into the slip, and Mary Beth leaped off the boat with the lines and tied the boat down. Neither Bill nor Rick said a word; Bill shook his head.

# CHAPTER EIGHTEEN

## Ithaca

Monday morning at nine a.m. at the office, Mike's intercom sounded.

"Mike, Gary Becker on line one," Miranda announced.

"Got it," Mike said, picking up the receiver.

"Hey, Gary."

"Hi, Michael. Look, I'm just giving you a heads-up. I'm presenting the Weber case to the Grand Jury this week. I'll be sending you a formal notice today. You think you'll have your client testify?"

"Have we ever, Gary?"

"Of course not, but I have to ask."

"I appreciate it. She won't be testifying."

"Got it."

***

Nine women and seven men made up the Grand Jury. Of the women, six were mothers with young children, two were retired senior citizens, and one was the mother of two college students and was going through a divorce. Two of the men were grandfathers, three had children in grammar and middle school, and two were single. Their education levels varied from high school grads, college grads, and PhDs teaching at the university. It was a well-diversified jury. Every one of them had heard of the Weber murder, yet every one of them swore under oath that knowing what they did from the newspapers, internet and local TV would not prevent them from rendering an impartial verdict.

They had been at work now for two weeks and were weary. Unlike a jury at trial, a Grand Jury would sit through presentations of several cases by the prosecution over several days, sometimes weeks, sometimes months.

*The case of The People of the State of New York v. Alina Weber* came up as their twelfth case. As is usual in Grand Jury Proceedings, the district attorney, and in this case, Gary Becker, DA, presented the evidence. There is no judge present, and defense attorneys are not allowed in the Grand Jury hearings unless the accused chooses to testify. So, Gary Becker had the show all to himself.

He first called Deputy John Wilson. Wilson was first on the murder scene with his partner, Samuel Irving. He testified about getting a call reporting a possible gunshot and going to the Weber residence to find the front door open, Weber dead on the couch, and Alina Weber alone in the house. Sam Irving testified about the scene, pretty much echoing what Wilson said. He left out the part where he threw up in the driveway.

Detective Cindy Maddison testified about her arrival at the scene with the CSI team and finding the gun under the mattress in the bedroom and three spent .38 shells. She also testified that the fingerprints—and the only fingerprints on the gun—were Alina Weber's. She confirmed Alina Weber had no license to carry a firearm and possessed it illegally.

FBI Special Agent Tom Foster testified, verifying the gun

found at the scene was indeed the murder weapon. The coroner testified as to the obvious cause of death: three gunshot wounds to the head.

After each witness's testimony, Becker entered the various written reports testified to by the witness into the record.

The indictment was handed down swiftly and unanimously: Criminal possession of a firearm and murder in the second degree.

Gary returned to his office on the third floor of the courthouse and called Michael.

"Got your indictment," Gary said with satisfaction.

"I figured," Michael said.

"So, you want to turn your client in, or you want me to have her picked up?"

"No, I'll bring her in."

"Good. Arraignment is tomorrow at nine, county court. Bring her right up to the jail."

*Here we go,* Mike thought. "Sure," he said.

Mike's next call was to Alina's court-appointed social worker, Jason.

"Gotta bring her in, Jay," Mike said matter-of-factly. "She's been indicted."

"Sorry to hear that," Jason said sincerely.

"Well, we were expecting it."

"Is she at the Center?"

"No. She's here."

"Really?" Mike said, surprised.

"She's staying at the Center. She's just here for a visit."

"Hm. OK. Well, can you bring her and meet me up at the jail? We have to turn her in. Arraignment's first thing in the morning."

Alina saw the change in Jason's face. "Jason? What is it,

Jason?" she asked, knowing it was nothing good.

Jason turned to her and raised his index finger as if to say, "Give me a minute."

"OK, Mike. We'll be right up."

***

Vince and Mary Beth got home Sunday evening.

Monday morning, Vince was alone at the lake cottage enjoying a cup of Starbucks coffee and reading the latest issue of Boating Magazine. Mary Beth had left the house early that morning for a consulting job meeting with the First Federal Savings and Loan Company on an embezzlement investigation for the bonding company that had paid the million-dollar claim. Vince's phone sounded.

"Dad."

"Hi, Mike."

"How was the boat?"

"Good! Pretty good. Actually took her out for a test run. The waves were a little choppy, but—"

"Dad. Alina's been indicted, and the arraignment's tomorrow morning."

"Shit!" Vince blurted. "Charge?"

"What do you think?"

"Murder two,"

"Bingo."

"Where is she?" Vince asked.

"She's on her way to the jail with Jason, and I'm in my car on my way up there."

"OK, I'll meet you at the jail."

"Dad. Dad. You're second chair, remember. I'll handle this."

Vince paused, realizing he needed to back off and let his son lead on this case. Not an easy thing for Vince.

"Right."

***

Mike arrived at the sheriff's office building first and was greeted by Detective Ken Ludlow.

"Where is she?" Ludlow said, skipping the usual courtesies.

"She's on her way," Mike said.

"She better be."

"Did I just tell you she's on her way? She's on her way," Mike said, returning Ken's tone.

"Eh, sorry, Mike. Didn't mean to come across that way. Got a lot going on."

"No problem, Ken. You always do. Ah. Here they are."

Ken was swift to take Alina from Jason's care. Two other deputies appeared from the back office and promptly moved to handcuff her. Despite how Jason had tried to prepare her for what was about to happen when the first deputy grasped her arm and shoved it behind her back to apply the first cuff, she blacked out and went limp. Jason rushed to her only to be shoved back by the second deputy, who lifted her until she stood on her own. As she was escorted through the doorway to the jail, she turned her head, and her eyes never left Jason's until the solid steel door slammed behind her, and she was gone.

"You can talk to her in about half an hour, Mike. Gotta process her. We'll let you in the attorney room when she's ready."

"Sure," Mike said, then turning to Jason. "Thanks, Jason. You can head home now. I'm going to talk to her alone. You can come back later tonight if you want."

"There's nothing I can do tonight, Mike, but I'll be in court in the morning."

"Good. See you then."

Deputy Wilson escorted Mike through security and back into the jail part of the sheriff's office—bare concrete floor, gray concrete block walls, stark florescent bulbs encased in steel grates, and inch-thick glass between the sergeant's cubicle and the hallway of jail cells. The attorney "conference" room was a concrete room a mere six feet by eight feet with one-inch bullet-proof glass windows on two sides and with a plastic table and two plastic chairs, one on each side of the table. There was no door knob on the inside of the steel door with wire-mesh glass.

Wilson unlocked the door and let Mike in.

"I'll bring her back now," Wilson said. "And I'll be just down the hall when you're done, so tap on the glass when you're done, and I'll let you out."

Mike entered, took a seat; Wilson shut the door and left to get Alina. A minute later, he reappeared with her, unlocked and opened the door. She was dressed in an orange jail jumpsuit, handcuffed behind her back, and her ankles shackled, her head drooped. She shuffled into the room.

"Take those off," Mike said, pointing to the handcuffs.

Wilson paused, then complied. She sat, her shackles clanking against the legs of the chair. Wilson turned, left the room and slammed the steel door shut behind him. He disappeared down the hallway.

Alina rubbed her wrists. "They were too tight," she said.

"I'm sorry about that," Mike said.

There was a long pause, and she made no eye contact with Mike.

"How are you doing?" Mike asked. The question immediately felt to him like the most ridiculous thing he could've asked.

She looked up and met his eyes. "OK."

"That's good."

"Why am I in jail?" She asked. "Jason told me some things, but I don't understand."

Mike took a deep breath, leaned forward and put his elbows on the table, clasping his hands. "Well, Alina. I'm afraid to tell you that you've now been formally charged with the killing of your husband."

Her eyes never left his. "I didn't kill my husband," she said flatly.

"OK, Alina. We're going to talk about this. But for now, I want you to know that I'm going to defend you, and I'm going to work hard for you. Tomorrow, you and I are going to court in front of the judge and I'm going to enter a plea of not guilty to all the charges."

"Then I can go?"

Mike took a minute to think.

"I'll be blunt, Alina. Since the charge is now murder, it's likely the judge will vacate the current release order and remand you."

"That's what Jason said, but I don't understand."

"OK. So basically, it means the judge is going to vacate your bail, release your bail, and keep you in jail until trial."

"So, I stay in jail," she said, tightening her lips.

"I'm afraid so. Until trial."

"And then?"

"That depends on if you're convicted or not. If you're not convicted, if you're acquitted, then you go free. If you're convicted, if you're convicted, you stay in jail."

"But I didn't do it, Michael."

"We have a lot of work to do, Alina."

There was a tap on the glass, and Alina was startled.

"If she wants dinner, she's gotta come now," Wilson said through the glass.

Mike gave the 'OK' sign.

"OK, Alina, you go have dinner, and I'll see you in the morning."

"And Jason?" she asked.

"He'll be there," Mike assured her.

"OK. Thank you."

***

The county courthouse was built in 1929 of stone and white Italian marble. It stood in its original state with the only exceptions: a modern heat and air-conditioning system, updated electric, plumbing and the addition of an elevator and ADA compliance, including a ramp to the side entrance and handicap-accessible restrooms. Otherwise, it was near-perfectly preserved. The main courtroom was upstairs and consumed a full three-quarters of the second floor. The ceilings were thirty feet high, with gold leaf trim, and the stained glass windows depicting ancient courtroom scenes and symbols of justice were floor-to-ceiling across one full wall. Six six-foot-diameter brass and crystal chandeliers hung from the ceiling. Three sections of ten rows of dark mahogany pews gave the courtroom a church-like appearance.

Today, more than half the pews were full of spectators, attorneys and local news reporters as Alina's indictment had quickly hit the internet. Mike met Vince and Jason in the hallway outside the courtroom entrance, and they went in together. Jason sat in the second pew; Mike and Vince sat in the front row reserved for attorneys. Mike looked across the room, and Gary Becker was already seated at the prosecutor's table—the one close to the jury box.

As they sat, the bailiff entered the courtroom from the door to the left of the judge's bench, pounded the door with the palm of his hand, and in a raised voice said, "All rise, the honorable Jennifer Britton presiding!"

He held the door open, and Judge Britton strode to the bench, her black robe flowing behind her. She stood behind the bench and scanned the room.

"You may be seated," she said as she sat. She turned to the court clerk already seated in front of and to the judge's right.

"First case, *The People of the State of New York versus Alina Weber,*" the clerk announced.

Mike and Vince left their seats, pushed through the gate in the bar, and sat at the defense table at the left of the courtroom, leaving an empty chair between them. A second bailiff escorted Alina through the door to the far side of the judge's bench and to Michael and Vince. She was dressed in the bright orange jail jumpsuit with TOMPKINS COUNTY JAIL written across the back, handcuffed in front of her, and her ankles shackled, with a chain running around her waist and attached to the shackles by her feet. She could only make tiny shuffles as she struggled to walk. Mike pulled the chair to his left, away from the defense table, and she sat, her chains and shackles making a clanking sound that echoed through the courtroom as they collided with the wooden chair.

Judge Britton took her time alternately scanning the computer screen and open paper file in front of her on the bench. Britton was fairly new to the job. She'd spent the majority of her legal career of twenty years as a criminal defense and family court attorney and children's advocate. This was in stark contrast to her predecessor, who had been an aggressive prosecutor before he took the bench. Although she would never admit it, her experience made her a compassionate, patient, respectful, and empathetic judge. None of this, however, affected her ability to sharply and precisely apply the law to the facts of every case before her to ensure a fair, and proper result in her courtroom. She was well respected by the attorneys and other judges in the state for her often profound, well-thought-out written opinions and judgments. Still, she was not a "defense lawyer's dream," nor a prosecutor's. She applied the facts of every case to the pertinent law and made her calls accordingly. If compassion was in order, as in a non-violent drug offense or cases involving minors, she was quick to make referrals to a social worker or treatment rather than a jailer.

"Mr. Becker?" Britton said.

Gary stood. "Your honor, we're here this morning in the case of the *People v. Weber.* The case has been indicted, and the indictment was handed down. The People state readiness for trial."

"You're not ready for trial," Vince said under his breath, "you always state readiness for trial, then let's just try the case right now."

Becker heard Vince, turned to him and shook his head.

Judge Britton pulled the indictment from the folder in front of her. "I have it." She turned to Vince and Mike. "Mr. DeMarco, you're defending Ms. Weber?"

Vince began to stand; Mike put a hand on Vince's shoulder and gently pressed him back down into his chair. Mike stood.

"Ah," the judge said. "So I have the young Mr. DeMarco today."

"Yes, Judge," Mike said. "We filed our Notice of Appearance this morning."

She flipped through the folder. "I don't see it. Wait." She scanned the computer screen. "OK. I've got it. You may proceed."

"Judge, we waive a reading of the charges, and we waive a

reading of my client's rights, and enter a plea of not guilty to the two charges in the indictment. We would like to be heard on bail."

Mike knew he had nowhere to go on bail but wanted to make the argument for Alina's sake.

"Bail?" Britton said, "Do I have a choice on bail?"

"Well, your honor, as the Court knows, Ms. Weber has posted an extraordinary amount of bail on a firearms charge, and I submit that bail be continued on the new charge since it's apparent that the amount of bail set was in anticipation of the murder charge coming. Why else would Judge Vance have posted a half-million-dollar bail on a gun charge?"

She turned to Gary, "Mr. Becker?"

Gary was smirking. "You asked if you have a choice, Judge. I suggest you do not. This is a second-degree murder charge, and no bail is available in this state for such a charge."

"Mr. DeMarco, although I appreciate your argument and your effort on behalf of your client, I have to agree with Mr. Becker. As such, I am vacating the release order of your client, releasing the bail, and remanding the defendant pending trial. My court clerk will advise you of a date for a conference if needed."

With that, she stuck the gavel. "We are adjourned." She stood and left the courtroom.

Jason sprinted up to the bar and tapped Alina on the shoulder. She turned.

"I'll be right up to see you, Alina."

She was confused, frightened, and in shock and said nothing. She turned to Michael.

"What happened?" She asked.

Mike's eyes met Vince's, and then he turned to Alina.

"Alina, the judge is not going to let you out of jail until trial."

"Yes, you said that to me before, so there is no chance?"

"No, Alina, not until trial, and then only if you're acquitted, found not guilty."

"OK, so when is my trial? Is it soon?" She asked.

Vince piped in. "We don't know, Alina. But Michael will be having conferences with the judge and Mr. Becker before there'll be a trial date set."

"Conferences?"

"Yes. Conferences to talk about the case, and we'll be filing motions to fight for you."

Alina just shook her head; too much for her to take in. The court bailiff appeared at the defense table, and she was swiftly escorted out of the courtroom, leaving Michael and Vince standing at the defense table.

In the back of the courtroom, "Nikolai Volkov" watched intently.

Mary Beth had finished her meeting at the bank with the bank president, and the bonding company claims adjuster and had rushed to the courthouse to watch the arraignment. She arrived a minute too late, just as people were shuffling out. Vince saw her out of the corner of his eye and went to her. She wore a navy blue Alexander McQueen single-button blazer pantsuit and matching stilettos. Heads turned.

"I guess I'm late," she said.

Vince gave her a quick kiss on the lips. "Yup. You missed the show."

Just as he said that he spotted Volkov in the crowd heading to the stairs. He was in a hurry, almost pushing people out of his way to get down the stairs.

"What are you looking at?" Mary Beth asked, noticing Vince's sudden distraction.

"That's Nikolai Volkov," Vince said. "That's the guy that posted Alina's bail."

She followed Vince's gaze and squinted her eyes as if to better focus. "Which one?" She asked, watching the crowd of people descending the steps.

Vince looked at her and back to the stairs; he was gone.

"Never mind; he's gone," Vince said.

For a split second, Mary Beth was confused. Then, it dawned on her as her memory came into focus.

"Vince, can we talk somewhere?" Mary Beth asked.

"Of course, why?"

She motioned to Mike, who was talking with Becker by the prosecution table.

"Him too," she said. "I'll meet you in the hallway."

By the look on her face, Vince knew this was something serious. Vince walked over to Mike and tapped him on the shoulder, interrupting Becker in mid-sentence.

"Sorry, Gary. I have to talk to Mike." Vince said.

"Sure."

Mary Beth had made her way out of the courtroom and down the hallway away from the people leaving. The three gathered at the end of the hall.

"Vince, Mike, I know Nikolai Volkov,"

"What?" Vince said, incredulous. "What do you mean?"

"I mean, I know Volkov. He was a target of the criminal investigation I was on In DC before I was reassigned up here...to you."

Without saying a word, Mike began running down the hallway to find him. Vince and Mary Beth followed. By the time they reached the sidewalk outside the courthouse, he was nowhere to be found.

They gathered by the steps.

"The bail!" Vince said. "He'll be going up to the jail to get his bail money since the judge released it."

"Right," Mike said.

"Well, that can't happen until the judge issues the order releasing it. That could take days, even weeks," Mary Beth said. "Did she release the bail?"

*God, she's good,* Vince thought. "She did."

"I think we need to put Ken Ludlow on notice," Mike said.

"And talk to Alina," Mary Beth added.

"Yup," Vince said. "We better have a talk with her. I'll meet you at the office, Mike."

"I'm done for the day," Mary Beth said, hinting.

"You're done for the day?" Vince asked.

"I'm done for the day," she repeated.

Mike understood. "Dad, I'm going back to the office. You head home."

"Ah," Vince said. "Old habits and all."

Mary Beth took him by the arm, and they left.

Jason had made his way back to the judge's chambers and approached Judge Britton's clerk, Marion.

"Could I see the judge for a minute?" Jason asked.

"I'll see," Marion said as she pressed the Intercom button on her phone. "Judge, Jason Valor would like a minute of your time."

"Sure. Send him back," Britton responded.

"You can go in," Marion said, pointing to the door to the Judge's office.

Jason turned the polished brass knob on the heavy oak door and entered the inner chamber. He took a second to take it in: twelve-foot tall ceiling, gold-trimmed crown molding, crystal chandelier, hunter green and gold-leaf wallpaper, and Judge

Britton sitting behind a massive ancient mahogany desk, still in her robe.

"Jason, what can I do for you?" She said, leaning back in her black leather chair.

"Well, Judge, it's actually quite personal," Jason said with a small smile.

"Oh, in that case," she said, as she walked around her desk toward him, "Let's sit over here." She pointed to a well-worn leather maroon couch and matching chair that Jason figured had to be at least fifty years old.

Jason sat in the chair; Judge Britton sat leisurely on the couch.

"What's on your mind, Jason?"

He took a moment to gather his thoughts. She smiled at him, and the smile encouraged him to go on.

"Well, Judge, I've decided to quit social work, so I won't be taking any more court assignments. I will finish out the Weber case."

Britton stroked her chin, "I see. Can you tell me why?" she asked, knowing full well how daunting and exhausting social work is, especially in the court system.

He took a deep breath, pondering how personal he should get. She looked him in the eyes with a tight-lipped smile. "Go on," she said.

"Honestly, I'm tired, stressed and frustrated. I don't know that I'm doing anybody any good."

"I see," Britton said, sitting up and leaning toward Jason. "It's a tough job you have."

"Ah, no tougher than yours, Judge."

"Oh, I don't know about that," she said. But I know what you have to deal with, Jason. People don't come into this system because they're living a dream; they come into the system because they're facing the worst problems in their lives: arrests, prison, custody battles, divorce, addiction, abuse, death, and we have to do what we can to sort it all out, to keep society running. And you're on the very front line, in the 'trenches.' I know, and I can only imagine how daunting it must be, day in and day out."

Jason was surprised by her understanding, kindness and empathy. He'd expected a rebuke or, at best, a quick acknowledgment and dismissal.

Britton went on, "You said you don't think you're doing anyone any good. Do you really believe that?"

"The Donner case," Jason replied.

Britton leaned back on the couch and crossed her arms in front of her. "Ah, yes. The Donner case. Fooled us all, didn't he?"

"Well, he took *me* for a fool," Jason admitted. "And I filed my report actually believing him. He lied; he used me; you went by my report and released him. I failed, and his wife suffered for it."

Britton nodded, a slight frown on her face. "I see, and I understand, Jason."

She wanted to console him, tell him it wasn't his fault, that Donner was a conman, but she knew it would do no good.

"So where do you go from here, Jason?"

"Well, like I said, I'll finish out the Weber case, then I'll fall back on my computer science minor in college."

"That's quite a switch," Britton said.

"It is," Jason said.

"Well, that explains why you were able to help us install the new electronic filing system this year. None of us could've done it without you," she said. She stood and reached her hand out to him. He stood, and they shook hands.

"I know you'll do great, Jason, and you were a great asset to us here."

"Thanks, Judge," he said as he left her chamber and closed the door behind him.

*It's a burn-out job,* Judge Britton thought to herself, *a burn-out job.*

# CHAPTER NINETEEN

As promised, Jason left the courthouse after the arraignment and headed up to the jail to see Alina and talk to her about providing her with medical and psychological care while at the jail. He stopped at a Starbucks, bought two coffees and arrived at the jail twenty minutes after Alina. When he got there, there was no one at the entrance lobby, so he knocked on the one-way glass to the side of the steel door entrance to the jail part of the sheriff's department. The sergeant responded through the intercom speaker.

"Hey, Jason. Who you here to see?"

"Alina Weber."

"OK. Have a seat; you're gonna have to wait a while; she's with her lawyer."

"DeMarco's already here?"

"DeMarco? No. Her other lawyer."

Jason's eyebrows raised. "Her other lawyer?"

"Yeah. Adams. Sean Adams. He was already here waiting for her when she got back from court. He's in there with her now."

"Hmm. I had no idea," Jason said. "Could you let her know I'll be up to see her tomorrow?"

"Sure, kid. Have a good day."

"You too, Sergeant," Jason said, shaking his head as he left, taking the coffee with him.

***

When the deputy told her her lawyer was in the attorney conference room waiting to see her, she expected to see Michael DeMarco or maybe Vince. As she was escorted down the hallway toward the room, she saw it was Sean Adams and froze in her tracks.

"What's wrong with you?" The deputy asked, "Come on, *move*."

The blood drained from Alina's face, and her knees weakened. The Deputy sensed it and lifted her under her arm, moving her forward down the hall. Realizing the state she was in, he offered to take her to the cell she was going to share with another young woman, a girl in her late teens arrested for cocaine possession. When they reached the door to the attorney conference room, the deputy pulled a key from his pocket and undid her shackles and cuffs. "You don't need these now. You're not going anywhere. You gonna be OK, you want to go to your cell?"

Alina took a deep breath. "No. I'm OK," she said in a soft tone. He unlocked the door, let her in and shut it behind her. She sat in the white plastic chair opposite Adams, her head held low, her eyes looking down to the table, her arms crossed tightly in front of her. She said nothing.

"Alina," Adams said softly.

No response.

"Alina," he tried again.

No response.

"Alina, are you alright?"

With her head still down, she slowly looked up at him. When their eyes met, Adam's heart began to race. He took deep breaths to control his spiking emotions. Alina could not bring herself to speak. A single tear trickled down her cheek.

"Alina, everything is going to be OK. I promise," Adams said.

She shook her head slowly. "How can you say this?"

"Because you're innocent."

A sardonic smile came across her face. "Of course I am. Who would know better?"

"Did you doubt my word?" he asked.

"Doubt your word? You mean when you told me you loved me so much, you'd do *anything* for me?"

"Yes."

"You killed for me?"

"Yes, because I love you."

"Oh my God, I never should have told you about what was going on in my marriage. I never should have written 'help me' on your pad."

"You didn't have to, Alina. I saw the bruises on your face."

"It did not go well, Sean."

"I know. I know, Alina."

She was now sitting erect and glaring him straight in the eyes, her arms still crossed tightly in front of her.

"Now look what you've done to me," she said.

"But you're free, Alina, free from him."

"This is not free!" She said roughly tugging on the front of her orange jail jumpsuit. "This is not free!"

A spike of panic shot through Adams.

"Alina! Alina! You'll be free! I promise you; you'll be free! Please don't...don't..."

"Don't what, Sean? Tell the truth?"

Adams sat back in his chair, tightened his lips and slowly nodded.

She glared at him, now cowering in his plastic chair.

They sat in silence for what seemed to Adams an eternity. Finally, he found the strength to speak up.

"I've found new evidence that will free you, Alina."

No response.

"I have proof, I mean information that will fix this."

She raised an eyebrow, a slight scowl on her face.

"And Alina, you have the best defense lawyers in the state. I've already contacted them with the information, and they're excited about it. You'll see! It's going to be all good. Please just be patient."

"What choice do I have?" she asked.

He knew exactly what choice she had, and it terrified him to the core of his soul.

"I love you, Alina."

She shook her head.

"I thought you loved me," he said.

"You left the gun," she said softly.

Adams jolted and looked out the glass to see if the deputy was standing by the door; he wasn't.

"Jesus, Alina!"

Adams leaned as close to Alina as he could and spoke in a whisper. "Alina, look. If you say it was me that shot Weber, I'll testify you agreed to it, and that's conspiracy to commit murder. You're as guilty of killing Weber as I am. You might better just let me work on this and get you out of it, understand?"

She stood. "Get out...get out now," she said.

She tapped on the glass, getting the deputy's attention and gesturing she wanted out. The deputy opened the door.

"You leave first, counselor, and have a good day," the deputy said.

He then escorted Alina to her new living quarters.

She sat on the thin-mattress metal cot in her cell, staring

blankly across the tiny gray cinderblock room. There was a second bunk, but no one else was in the cell. She crouched over, her elbows on her knees, her head in her hands, and wept—not in sadness, but in unfettered anger. She had planned to tell Michael that it was Adams who shot her husband, but that was no longer possible. Adams had made it perfectly clear that if she did, he would make sure she, too, would spend the rest of her life in prison.

***

Adams was sick to his stomach. He'd gone to see Alina to tell her how information about Weber's ex-wife and the life insurance policy would exonerate her and point the blame on the ex, but instead, it all blew up in his face. Try as he might, he couldn't remember what he did with the gun after he shot Weber. Leaving it there was a fatal mistake for Alina, and probably eventually for him too. He had no faith that under the pressure, Alina wouldn't eventually turn on him to save herself.

He had to pull it together now because he was to meet with Helen Stinson Weber in twenty minutes at his office.

***

When he got to his office, Helen was already there waiting for him in the hallway. She was much taller than he'd expected, knowing how short Dr. Weber was. She appeared reserved, dressed in a brown tweed blazer and skirt.

"Ms. Weber?" Adams asked.

"Stinson. Helen Stinson. Mr. Adams?"

"I am. Come in, please."

He unlocked the solid oak door, opened it, and gestured for her to enter. She first noticed the grime around the doorknob, then was surprised at how small the one-room office was.

"Have a seat," Adams said, pointing to one of the vinyl and metal chairs in front of his desk. "I see you've brought papers with you."

"I brought the policy. Here," she said, handing it to him across the desk.

Adams took it, set it on the desk, and flipped through the first few pages of the booklet.

"What kind of policy is it?" she asked.

"Well, it's a life insurance policy," he responded, puzzled by the question.

She raised an eyebrow. "I know that," she said flatly. "What *kind* of life insurance policy?"

Adams remembered back to his law school days and couldn't remember there being any courses offered on insurance. *It's just a contract,* he thought to himself, trying to gather some confidence. He closed the booklet.

"It's the 'Presidential' policy," he said, reading the large blue letters on the cover.

"Yes. I can read that. Can I cash it in?"

"Well, we'll have to file a claim on the policy."

"No, no. That's not what I mean. Does the policy have cash value?"

Adams flipped through the pages, buying time to think; she was confusing him, and the more he tried to read, the more thoughts of Alina and the disastrous meeting he'd just had with her filled his head. He'd cut to the chase.

"It doesn't matter what kind of policy it is; he's dead, and if the policy is in force, we file the claim, and you get the money."

His sudden, abrupt demeanor took her aback. She leaned

forward in her chair. "Here," she said, sliding a copy of Weber's death certificate across Adams' desk.

It took Adams by surprise.

"Where did you get this?" he asked.

"The funeral home," she said. "I told them I was his sister."

*For chrissakes,* he thought, *Everything is just getting worse by the minute.*

Adams looked at, but couldn't bring himself to pick it up, and left it where she put it.

"You'll need that, won't you?" she asked.

"Yes. Yes. To file the claim."

She reached into her purse and pulled out a document folded in three with a blue backer. "Here's his Will. He left everything to me."

"No, no, no," Adams said, shaking his head. "You were divorced, and he remarried."

"Oh, I doubt he left anything to that girl," she said. "You know what she is."

"No, I don't. I don't know what you mean."

"Sure you do," she said, placing the Will on top of Weber's death certificate. "You were his lawyer; you still are."

"But the judgment of divorce would exclude you from his estate in general."

"Well, look, Mr. Adams, you have the life insurance policy. He never changed that. And as far as the divorce, he never did what he was supposed to do in the separation agreement. The house is still in both our names, and I never got half his retirement account that was ordered by the court. So maybe you have a lot of work to do."

A wave of nausea rolled over Adams, and his vision tunneled.

"Just a minute," he said as he stood. He went into the small bathroom of his office, closed the door behind him and splashed cold water on his face. When he returned to his desk, Helen Weber was gone; the death certificate, Will and insurance policy unmoved. He dropped into his chair, his elbows on the desk, his head in his hands and stared at the pile of papers in front of him. His mind raced. He never should've told Alina he might testify against her if she implicated him. That wasn't fair to her, but he panicked at the way she seemed to turn on him at the jail. Now, Helen Weber Stinson, with her hand out for whatever she could get.

*And who do I represent?* he thought, *Dead man Weber? Helen Weber? Weber's widow Alina? Oh my God, the Weber estate? What a mess I'm in! What have I done? She doesn't even love me! Oh my fucking God! She doesn't even like me!*

He sat paralyzed for a minute, or an hour; he couldn't tell; he'd lost track of time. Finally, he gathered his wits enough to open Weber's Will. Sure enough, he left everything to his "loving wife, Helen," no contingent beneficiary. He flipped to the last page, and he'd also named her executor of the Will. He slowly folded it and set it to the side. He took a deep breath, and pulled the death certificate toward him, leaving it laying on the desk. His eyes went straight to the cause of death: "Cranial GSW x 3, death instantaneous." Three? Three gunshots? He had no recollection whatsoever of how many times he'd pulled the trigger.

In his few years in practice, he'd almost exclusively done medical malpractice defense cases and those he did under tight supervision from his assigned senior partner in Syracuse. This was all over his head, and he knew he was drowning. The thought of talking to his assigned senior partner entered his head, and he tried to imagine how that conversation might go. He couldn't imagine how to even start that conversation. But that partner was his supervisor because all he did was malpractice defense anyway. The feeling of complete isolation and the urge to talk to somebody—anybody—was overwhelming.

***

The sun was big and red and appeared to be sitting on top of the hill on the far side of the lake as it set into the West. Vince and Mary Beth sat at the dining table side-by-side, enjoying the view and each other over dinner and red wine.

"You look tired, Vince," Mary Beth said in a warm, caring voice.

"Who me? Tired? Naw." Vince said. "Well, OK, maybe just a little."

"Are you worried about the Weber case?"

Vince took a sip of wine and time to think.

"I don't know, Mary Beth, I don't know."

"You have faith in Michael?"

"That's a silly question. Of course, I do."

"Mhm."

"I absolutely do, but I have to tell you, there's something dark about this case."

"Dark? It's a murder case, Vince! Of course, there's something dark about it."

"No. No. I've defended a lot of murder cases. There's something we're not getting here. It's like when the ocean draws back just before a tsunami rushes in, like a rumbling under the earth before an earthquake. I don't know."

Mary Beth drew a deep breath, and in her mind, she agreed with his take on the case. *There is something very dark here,* she thought, *very dark.*

"Are you worried for Mike?"

"I'm always worried for Mike."

Mary Beth smiled, kissed Vince on the cheek and sipped her wine.

The sun was just a sliver of red atop the hill across the lake now and cast a crimson glow across the water.

"Look at that," Vince said.

"Amazing, Vince. Just amazing."

Vince's phone sounded.

"I thought we agreed to turn those off," Mary Beth said, putting her glass of wine down on the table.

Vince froze and let it ring until it stopped, then picked it up to turn it off. The screen showed, "Missed call Michael."

*Damn,* he thought.

"Was it Michael?" Mary Beth asked.

"It was," Vince said, turning off the phone.

"I see," she said, turning her attention to the food on her plate, and counting silently to herself, *one...two...three...*"

"Yeah. It was Mike, so..."

"Oh, Vince, go ahead and call him; it's OK."

Vince picked up his phone, restarted it and tapped Mike's number.

"Michael."

"Hi, Dad."

"What's up?"

"Do you know Sean Adams?"

"Sounds familiar," Vince said.

"Sean Adams, the attorney."

"OK, Sean Adams, the attorney. The guy with the 'case-breaker life insurance policy.' So?" Vince asked.

"He called the office and left a message with Miranda looking for you," Mike said.

"Me? Did you talk to him?"

"Nope. He said he'd only talk to you, and it was about Alina, and it was urgent."

"Well, I'm not going to call him now! I'm in the middle of dinner with Mary Beth."

"I don't blame you, Dad; I just thought you should know. If you want, I'll text you his number."

"OK, Son. I appreciate it. But I'm not calling him tonight."

"Alright, have a good night, Dad."

"You too, Son."

Vince clicked off his phone, smiled at Mary Beth, and picked up his fork.

"Oh for heaven's sake, Vince, go ahead," she said, smiling and shaking her head.

Vince tapped on the number Michael had texted him. Adams picked up on the first ring.

"Mr. Adams, this is Vince DeMarco. You were looking for me?"

"Oh! Mr. DeMarco. Thank you for getting back to me. We need to talk."

"We are talking. What can I do for you?"

"Not over the phone. Can we meet?"

"What's this about?" Vince asked, starting to lose his patience for losing his precious time with Mary Beth.

"It's about Alina Weber, your case. Can we meet? I'll meet you anywhere you want."

"It's not my case; it's Michael's. You need to talk to him."

"I appreciate that, but it's you I need to speak with," Adams insisted. "I need advice; I need your experience."

"Alright. I can see you tomorrow. I can see you at my office midmorning if you want."

Adams took time to think.

"Mr. DeMarco, this is personal, very personal. Would you mind meeting me at my office?"

"You have an office?"

"Yeah. In the Circles building on the second floor."

"I didn't know that building was still in use," Vince said.

"Second floor, room 201."

"Fine. I'll be there at ten," Vince said, then clicked off the call.

"Hmm," Vince said, turning to Mary Beth. "He needs my 'experience.'"

"Well, you've got plenty of that!" she said with a big smile.

"Very funny, Mary Beth, very funny."

"You're meeting him at his office, not yours?"

"Yeah. I guess so," Vince said.

"Why?"

"Curious, I guess. Just curious. He says it's very personal."

"Aren't all your cases? I mean, isn't it all personal?"

"I'll meet him at his office."

"Well, actually, Vince. That's probably a good idea. Every time you step into your office, you get dragged back into the cases you've given Michael to take over."

"Can't help it, Mary Beth."

"Oh I know. Vince; I know."

# CHAPTER TWENTY

The next morning at nine o'clock, FBI Special Agent Tom Foster, Detective Ken Ludlow, and District Attorney Gary Becker met in the conference room in the DA's offices on the third floor of the courthouse.

"No results on the GSR swab yet. They told me they're running the labs a second time," Foster reported.

"That's not good news," Becker said. "She was the only person at the murder scene, and she was processed fast. She didn't even have a chance to shower."

"Well, we did everything right," Ludlow piped in. "And no, she had no opportunity to shower or clean up. The gun was still warm when CSI found it in the bed."

"Did she change her clothes?" Becker asked.

"Don't think so. We searched the place thoroughly just for that purpose and didn't find any clothes that had been tossed aside or hidden, just clean clothes in the drawers and closet, and I doubt she would've had time to change anyway. Deputy Wilson was there fast," Ludlow said.

"Besides, Gary, her hands, fingers, arms, and neck were swabbed, so even if she had managed to change clothes, they would've picked up GSR, maybe blood spatter if there was any. Maybe they'll pick some up on the second lab run," Foster said. "We ran the clothes she had on when the deputies went to the scene to the lab and not a drop of blood spatter on any of them."

Becker felt his face heat. "Great! That's all we need! Give DeMarco a place to go for reasonable doubt!"

"I got more," Foster said, noticing Becker's face turning red. "The gun."

"OK, the gun," Becker said. "What about it?"

"Remember, we got a hit on the serial number. Gun was sold new out of Fort Lauderdale eight years ago. Legit gun dealer, legit registered buyer. From there, it was sold three more times, all in the Fort Lauderdale area, then once in Miami. Then stolen from a retired school teacher in Ocala, turned up in a drug deal gone wrong in South Miami." Foster said.

"Of course, I remember," Becker said.

"Well, we traced down everybody involved in that drug deal."

"And?"

"And, it was a 'pound of flesh' shooting."

Ken chuckled.

"A pound of flesh shooting? And what's so funny?" Becker said, glaring at Ludlow.

"You want to tell him, or should I?" Foster asked Ken.

"Oh, let me, please," Ludlow said, turning to Becker. "It's when a drug deal goes bad, and the dealer shoots his customer in the cheek of his ass. See? A pound of flesh."

"Yup," Foster said. "Just to send a message."

"I see," Becker said. "And the bullet?"

Foster smiled. "They had the bullet, and we got the bullet; sent it to the lab at the National Tracing Center, they ran it through the National Integrated Ballistic Information Network, and lo and behold a match."

Gary sat up in his chair. "And?"

"Same gun shot that pound of flesh, shot Weber," Foster said.

"OK, so where's that get us?" Gary asked.

"To the shooter in Miami," Ludlow said as he pulled an envelope out of his jacket pocket. He slid a sheet of paper out of the envelope. "James Curtis, street name Big Red."

"Big Red?" Becker asked.

"Well, here's his mug shot," Ludlow said, "big guy, red hair, red beard so ... Big Red."

"All right," Becker said, "we have the gun in Curtis' hand; now, how do we get it into Alina Weber's hand?"

"Don't know," Ludlow said.

"Don't know? What about the FBI?" Becker asked, turning to Foster.

"There's a little problem there, Gary," Foster said.

"And?" Gary prompted.

"The bullet was retrieved, but the gun was never found, and since the case wasn't prosecuted, it all just dropped there. No conviction on the shooting. Charged with assault with a deadly weapon, but apparently, the so-called victim refused to testify and disappeared. The charge was dropped."

"The gun's here now," Gary said, controlling his temper, "and it just killed somebody. We know where 'Big Red' is?"

"Well, looks like we can place Curtis as close as Syracuse," Foster said.

"Right," Ludlow confirmed. "Ran his criminal record. Ran his DMV and one ticket on route 81 near Syracuse, speeding eighty-five in a sixty-five zone."

Gary smiled. "That gets us mighty close. Did you pull up the entire docket?"

"No, just the driver's license abstract," Ken said.

"Hell, I can pull it up," Gary said, opening his laptop on the conference table. He tapped in a few keys and squinted at the screen. "You're not going to believe this," he said. "Guess what."

"What?" Ludlow asked.

"Big Red had counsel for his speeding ticket."

"No shit. For a speeding ticket?" Ludlow said. "Who?"

"Sean Adams. Adams filed a notice of appearance with the court and got the ticket reduced to seventy in a sixty-five."

"Adams?" Foster asked.

"He was Weber's malpractice attorney! One of our local boys."

"Oh my God," Ludlow said, suddenly very serious. "He was up at the jail visiting Alina Weber."

They all sat in silence for a moment, pondering, until Foster spoke up.

"So, let me get this straight," Becker said. "Gun was sold new out of Fort Lauderdale eight years ago. Legit gun dealer, legit registered buyer. From there, it was sold three more times, all in the Fort Lauderdale area, then once in Miami. Then stolen from a retired school teacher in Ocala, turned up in a drug deal gone wrong in South Miami, disappears for years, ends up the murder weapon in the Weber case, and this Adams guy happened to represent Big Red, the last person who might've had the murder gun for a speeding ticket up here?"

"And you say Adams was Weber's malpractice attorney," Foster added.

"And he's been up at the jail with Alina Weber," Ken said.

"You want to pick him up, Ken, or do you want me to? A gun crossing state lines is a federal gig, not local."

"Pick up who?" Gary asked.

"Adams."

"Whoa, whoa, whoa," Gary chimed in. "Slow down. Pick him up on what? Representing a drug dealer on a speeding ticket? The most we can do is ask him to come in and meet with us voluntarily. What we have so far is he represented a guy that used the murder gun in this case in a shooting in Miami, and the gun turns up at the Weber house."

Ludlow chucked. "Well, it doesn't take a Harvard lawyer to see that the gun had to have gone from Curtis to Adams to Alina Weber."

"No," Becker said, "We don't have anything to link Adams to the gun itself, and until I have probable cause for anything on Adams, we can't bring him in unless he's willing. And, by the way, I think Harvard lawyers are way overrated."

"Never actually met one," Ken said with a shrug.

"Adams won't talk to us anyway. Attorney-client privilege. No matter what, he can pull that on us," Gary said, slapping his hand on the conference table. "And damn it, no gunshot residue on Alina Weber yet! We still don't have that fucking gun in her hand!"

"We have her fingerprints all over the gun, and nobody else's," Ken said.

"Without gunshot residue, all that means is that she picked up the gun; it doesn't prove she pulled the trigger," Foster said.

"I know, I know," Gary said. "Well, I think you should both meet with Adams." He turned to Ken, "Ken, you get a hold of him and see if he'll cooperate. Then you both meet with him."

"I agree," Foster said. "Let's give that a try. Still, attorney-client privilege and all..."

"OK," Ken said. I gotta get back up to my office for a staff meeting. I'll call him and let you know what happens."

"Gotta head back to Binghamton anyway," Foster said. "You got my number."

"The sooner, the better," Gary said.

Foster and Ludlow stood to leave, and Foster stopped at the doorway.

"And, I'm going to track down our 'Mr. Big Red,'" Foster said.

"Maybe see what he has to say about the gun and Adams."

"Why would he talk to the FBI?" Ken asked.

"Oh, we have our ways," Foster said with a wink.

***

The Circles Building was next to the courthouse, separated by only a few parking spaces between the buildings. If the building could talk, it would tell of some of the most incredible, dramatic, chilling, life-and-death, frightening, earth-shattering, life-altering cases one could imagine. For over one-hundred-fifty years, the ancient building housed the most dynamic, experienced, and powerful attorneys in the state. Now mostly empty, it housed a handful of attorneys who rented rooms by the month in the upper floors for the convenience of being next door to the county courthouse.

At ten o'clock, dressed in blue jeans, a black button-down shirt, a light black leather jacket and Docker boat shoes, Vince made his way up the stairs and down the hallway to room 201. He knocked, knowing there would be no reception area, just a one-room office. The door swung open before he could knock a second time.

"Thanks for coming, Mr. DeMarco," Adams said, shaking his hand.

Vince was shocked at how cold and wet Evan's hand was and how pale his face appeared. Adams was wearing blue slacks, a white shirt, but no tie, his hair disheveled. There were papers scattered in no particular order across his desk, and Vince noticed Evan's suit jacket was crumpled on the floor to the left of it. This was something more than just 'personal' Vince realized. There was real trouble looming here.

"Have a seat," Adams said, motioning to one of the client

chairs; he walked around Vince and sat at his desk.

Vince said nothing; just waited for Adams to open the conversation. Adams nervously shuffled through the papers on his desk, trying to organize them into to neat piles, as Vince watched.

"Sorry about the mess," Adams said, flashing Vince a forced smile.

Vince said nothing.

"OK. Well. You're probably wondering why I wanted to see you," Adams said, folding his hands on the desk in front of him and trying to compose himself.

"Seemed important," Vince said. "You said, 'personal, very personal.'"

"Can we consider this meeting attorney-client privileged?" Adams asked.

"Hmm. That would depend on a lot of things, now wouldn't it," Vince said. "For instance, are you consulting with me on a personal matter, or a case you're working on, or are you asking for some general, maybe procedural advice, or is this just a friendly chat?"

"All the above," Adams answered without a pause. "All the above."

"I'm not going to promise you anything until you give me some idea of what this is all about, Mr. Adams."

"Sean. Please call me Sean."

"Alright, and you call me Vince. Now, give me an idea of what is going on here."

Just then, the phone on Evan's desk rang. He looked at it, and the screen showed: "Tompkins County Sheriff's Department."

"You need to get that? I can step out," Vince said.

"No no no. It's OK," Adams said, feeling his chest tighten. He stared at the phone until it stopped ringing.

Vince's better judgment was telling him this was all a very bad idea, and the thought of excusing himself and leaving crossed his mind, but then he knew Adams had told him this had something to do with Alina, and he felt compelled to sit and listen.

"I can tell you this, Sean," Vince said. "I'll sit here and listen to you, but my only interest is our client, Alina Weber. If you have something to say, good or bad, that affects her, there will be no attorney-client privilege for you in this meeting. Do you understand?"

"I do," Adams said, his voice constricted.

"Alright, then. Talk to me."

Adams rustled through the pile of papers on his desk and, without a word, handed the Presidential life insurance policy to Vince. Vince took it, looked Adams in the eye, and then opened the policy to the declarations page. It showed the insured Dr. Raymond J. Weber, M.D., Helen Stinson Weber beneficiary, and the face amount of five million dollars. It was a term policy, which means it has no cash value and pays nothing unless the insured dies.

Vince looked back at Adams. "So?"

Adams had expected a different reaction from Vince—a eureka moment—realizing he'd just handed Vince evidence of Helen Weber's motive to kill her ex-husband—documentary proof they could use to instill reasonable doubt with a jury that Alina had killed him, that Helen Weber was the murderer.

Vince handed the policy back to Adams. "Is there anything else?" he asked.

Adams cleared his voice. "Well, you see, this provides Weber's ex-wife with the real motive for the murder. I think any jury would see this and think, 'Five million bucks!' Of course, she'd kill her ex-husband for that."

"I see," Vince said calmly. "Have you ever tried a murder case?"

"Uh. No," Adams admitted.

"Have you ever tried a felony case?"

"No, I haven't."

"Have you ever done a jury trial?" Vince asked.

"Well, no, my senior partner does the trials. I do mostly the discovery, depositions, pre-trial stuff," Adams admitted.

"I see," Vince said. "But this policy, this policy would exonerate Alina?"

"Just reasonable doubt. It's just to raise a reasonable doubt," Adams said.

"OK, Sean. Anything else?

Adams was frustrated Vince wasn't understanding.

"The policy is in force. Weber's ex has been paying the premiums all along, even after the divorce."

"I see," Vince said, stroking his chin. "Anything else?"

Adams was rubbing his forehead now.

"Here," he said. "Weber's Will."

It took Vince only two seconds to read through it.

"I see. But this was pre-divorce, yes?" Vince said.

"Yes."

"OK. So this means nothing," Vince said flatly, handing it back to Adams.

"There's more. Weber's house is still in his name, and his ex's name, and his retirement was never divided according to the judgment of divorce. So, he still owed her that. So if she killed him, she would get the house, at least," Adams explained.

Vince held back the torpedo question until now.

"Who do you represent here, Sean?"

Sean took a deep breath to loosen his chest, "I don't know, Vince. In God's name, I just don't know."

Vince's lips tightened, and he slowly nodded. "You were Weber's malpractice attorney, yes?"

"I was."

"And you have his Will and life insurance policy."

"I do."

"So, I can assume you're now representing his estate?"

"Helen Weber brought me the Will and life policy."

"So, what are your intentions, Sean?"

"Look. All I wanted to do was to help Alina. She didn't kill Weber."

"Well, that's encouraging," Vince said, his temper beginning to rise. "Look, Sean. What you've done doesn't help Alina. Quite the opposite. Bringing this all out into the open could very well put a nail in her coffin."

Adams gasped. "What?" What? How..."

Until this moment, Vince had chosen not to bother to school Adams, but now he couldn't resist. In his ignorance, or maybe naivety, Adams was giving DA Gary Becker a powerful weapon.

"Here's how it really works, Sean. In New York, a judgment of divorce automatically disqualifies an ex-spouse from both collecting on a life policy even if she's still named beneficiary, as well as disqualifying her as the executor of his Will."

"OK, So Helen Weber won't collect the five million. So that's the end of that, right?" Adams asked.

"Oh, worse, Sean, much worse. You see, unless there's a contingent beneficiary named in the life policy, or Weber put a clause in his divorce that despite the divorce, Helen Weber would remain a beneficiary, the insurance money goes into Dr. Weber's estate. Now, unless there's a new Will—post-divorce—naming anyone else to inherit, his entire estate—including the five million from the life policy—goes to—"

"Alina," Adams said, finishing the sentence. "Oh my God."

“Yup, five million motives. Anything else?” Vince asked.

Adams put his head in his hands and began to weep.

Vince quietly got up, shaking his head, and left the office to go see Michael and tell him what had happened.

Evan’s phone sounded again: “Tompkins County Sheriff’s Office.”

# CHAPTER TWENTY-ONE

It was a hazy day with a wisp of fog floating across the lake with a gentle southern breeze. Mary Beth fought the urge as she sipped her coffee, looking out over the water.

*Old habits are hard to break*, Mary Beth thought to herself as she finished her cup of coffee, succumbed, and dialed the United States Attorney's Office number.

"Mary Beth Allison! My God, how are you?" Sandy Mitchell, Assistant United States Attorney said. "It's been a long time!"

"I'm good, Sandy! How are you?"

"Oh, you know, living the dream. So how are things up in quaint little Ithaca, New York?"

"Not bad, Sandy."

"You miss the game?"

"The FBI? Not really. I've got a good thing going up here."

"So I've heard. How is Vince?"

"He's good. He's trying to take things a little easy these days."

"Right. He's, what, sixty—"

"Anyway, Sandy. I need to talk to you about a case you and I had in D.C."

"Sure, Mary Beth. Whatever you need. Which case? I think we did about fifty."

"Sixty-four, actually," Mary Beth said.

"Anyway, sure. Which case?"

"Nicolai Volkov."

"Volkov. Hmm. Refresh my memory on that one."

"Sure. Boat sank in the Potomac, three dead or missing, one Naval Academy Cadet."

"Oh God, yes, I remember that one. Homeland Security, the Naval Criminal Investigative Service, The Naval Academy. You were the lead agent."

"I was for a while until I got transferred up here."

"Right. I heard about that one, too. Good job, Mary Beth. They really needed you up there. We're still talking about that one."

"So, What about Volkov?" Sandy asked.

"Well, since I was transferred, I don't know how the case in DC ended, or even if it ended at all."

"But what brings him up after all this time, Mary Beth?"

"Was he ever found? Last I knew, he was MIA when I left."

"I can't tell you much. And Mary Beth, you're not an FBI agent now, so..."

"I understand, Sandy, and I don't mean to impose on our friendship. But let me at least give *you* some information."

"You can tell me anything you want, Mary Beth. Go ahead."

"OK, so it appears Volkov, or someone calling himself Nicolai Volkov, has made an appearance up here."

"Up there? Ithaca?"

"Yes. And since it's all a matter of public record, I can tell you this. I'm pretty sure it was him who appeared in court at an arraignment of a Russian national first charged with illegal possession of a firearm and then indicted for murder. He posted a half-million dollars bail for her."

"So, he surfaced up there? Unbelievable," Sandy said.

"Not so fast," Mary Beth said. "He's calling himself Nicolai Volkov, at least that's what Vince told me, But I only got a glance at him. When I saw him in the courtroom for the arraignment on the indictment, I couldn't tell."

"Could be just a coincidence? I don't know that the name isn't a common one in Russia, you know, like John Smith?"

"Could be," Mary Beth said, "but he reached out to Michael DeMarco, Vince's son who's defending the case, and said he was from the Russian Embassy."

"Russian Embassy? Oh, please, I truly doubt that. If I can find some time, maybe I'll pull up the case and call you back," Sandy said.

"OK. Thank's Sandy."

They hung up, and Mary Beth went to pour herself another cup of coffee when the phone rang. It was Sandy.

"Wow. That was quick," Mary Beth said jokingly.

"No, really, Mary Beth. It only took me a couple of minutes. Look, the name Nicolai Volkov, or Nicolay Volkov, is a very common Russian name. In fact, Nicolai Volkov was a famous Russian actor who died in 1985. A lot of kids were named after him, so maybe you ran into another 'Russian actor' up there."

"Or maybe you and I ran into another 'Russian actor' back then in DC," Mary Beth quipped.

Sandy didn't appreciate that turn; it put the brakes on the conversation.

"Well, I've got some thinking to do," Sandy said. "I'll get back to you if I need any more information."

"I'll let you know if anything comes up, Sandy, and thank you for taking the time to talk."

"Sure. Just when I thought I couldn't possibly have more work to do, you call with this!" Sandy said with a chuckle.

"And one more thing," Mary Beth said.

"Yes?"

"Does the name, John Holdings or Holding, mean anything to you?"

Sandy took time to think.

"Holdings, John Holdings?"

"Yeah."

"Nothing comes to mind, Mary Beth. Why?"

"Not sure. Just a hunch."

"You can take a girl out of the FBI, but you can't take the FBI out of the girl, huh, Mary Beth?"

They shared a light chuckle.

"Guess so," Mary Beth admitted.

"Well, take care, Mary Beth and be careful."

"You too, Sandy."

"Oh, and say hello to Vince for me."

"Roger that."

***

John Holdings showed up at the Tompkins County Sheriff's Department civil window with the receipt for the bail posted for Alina and slid it across the counter, through the slot at the base of the bullet-proof plate glass to civil deputy Susan Alling.

"I'm here to get my bail back," he said to the deputy behind the glass.

"And you are who?" she asked.

"Nicholai Volkov. I paid the bail for Alina Weber."

Deputy Alling picked up the receipt, studied it for a minute, looked up at Holdings and said, "ID? You got ID?"

Holdings pulled a passport out of his inside jacket pocket and slid it over to her. She opened it to the first page.

"Volkov? Nicholai Volkov?" she asked.

"Yes."

"Just a minute," she said, then turned her back to Holdings and left into the back office.

Ten minutes later, she returned.

"No release order yet," she said.

"Release order?" he asked.

"I said, no release order yet. The judge hasn't released the bail yet," she repeated.

"Well, when do I get my money back?"

"Six to eight weeks."

"What?"

"I said, six to eight weeks! The money isn't here. Is that what you think? You think we keep the money here?"

"Well, no, but—"

"It'll be mailed to you from the finance department."

"Mailed?"

"Mailed to the address you gave when you posted it."

"Mailed?" Holdings said.

No address had been written on the bail receipt, just his name, the date, the amount paid, and that it was bail for Alina Weber. He had no idea what fake address he'd given when he paid the bail.

"Yeah. Can I make sure you have the right address? Can you tell me what address you have?" Holding asked.

Deputy Alling became even more suspicious than she was when the bail was first posted. She tapped her pen against the counter as she thought.

"Can't do that," she said, testing him.

"Well, I just want to make sure, you know, it gets to me at the right address."

"Why don't you tell me what address you gave, and I'll tell you if that's the address we have?"

Holdings felt his chest tighten and knew he was being tested, and this was a test he could not fail. It had been his responsibility to spring Alina, and he was the one who had made all the connections and arrangements to get her out on bail and move the cash. And he would be the only one to pay the price if this all went wrong.

"I … I have to go. I have an appointment," Holdings blurted as he turned and left.

*That's what I thought,* Alling thought to herself as she watched him sprint out the door.

The passport was still on the counter in front of her.

She buzzed Detective Ludlow on the intercom.

"Hey Ken, I gotta talk to you," she said.

"Come on back," Detective Ludlow responded.

She snatched up the passport and headed back to his office.

Ken was curious as to why the civil division deputy needed to see him, the detective head of the criminal investigations' unit. He sat leaned back in his chair, his arms crossed in front of him when she came into his office.

"Detective," she said, greeting him.

"Deputy. How are you?"

"I'm good, Ken," she said, dropping the passport on his desk.

He unraveled his arms, leaned forward, and picked it up.

"Nicholai Volkov," he said as he opened it and scanned the pages.

"The guy came in wanting Alina Weber's bail back. Something fishy about him; couldn't give me his own address."

"I see," he said as he slid his glasses down his nose for a better look at the document.

"Good grief! This doesn't even pass a level-one inspection."

"What's that?" Alling asked.

"It means this thing doesn't even pass an eyes-only exam."

"How's that?"

"Well, look. I'm no expert at all but look at the photo. What's wrong with it?" Ludlow asked.

"Uh, I don't know. Looks okay to me. And that's the guy that was here today looking for the bail money."

"Look again. He's smiling!"

"So?"

"You're not supposed to smile for a passport photo."

"I didn't know that; I've never had a passport."

"Well, other than that—and who knows if I'm right, I really can't tell you anymore. Just that it strikes me as suspicious."

"Welcome to my world," Alling said. "I'm going to leave it with you. It's over my pay grade."

"Good job, deputy. I'll take it from here."

"Thanks, Ken. I'll go back to driving my patrol desk."

"Very funny," he said.

"Haha."

"You need to get out more, Alling."

***

Ken waited for Deputy Alling to leave his office, picked up the phone and dialed Tom Foster, FBI Special Agent.

# CHAPTER TWENTY-TWO

**Washington, D.C.**

As soon as she hung up with Mary Beth, Sandy Mitchell, Assistant United States Attorney, sat motionless at her desk. Thoughts raced through her head. She had vague memories of the investigation into the fatal sinking of the boat on the Potomac Mary Beth had been working on when she was transferred to Upstate New York. That murder-for-hire conspiracy case led her to eventually fall in love with Vince DeMarco, the man she'd been assigned to protect. As that thought passed through her mind, she was surprised to feel a pang of jealousy over Mary Beth's new life. She snapped out of it and began tapping away at her computer to pull up the Volkov case.

What Mary Beth never knew was that when Volkov's 1951 Chris-Craft Riviera classic wooden thirty-foot mahogany speed boat was salvaged from the bottom of the Potomac River, at only sixty feet of depth, human remains were found ensnared in the wreckage. There was ample DNA material found with the scant human remains, but testing and analysis were not ordered because it was classified as an accident, and the tissue samples were destroyed.

Samples were never submitted to the Federal DNA Database Unit (FDDU), the organization law enforcement agencies use in investigations looking for hit confirmations against people whose profiles are in the National DNA Index System (NDIS). There would be no hits for anyone convicted of, arrested for, or facing charges of crimes or anyone else that would've submitted to DNA sampling for any other reason, including jobs that required it for background checks or contested paternity cases. Nor would there be any DNA samples collected at unsolved crime scenes anywhere in the country that matched. The tissue samples were labeled John Doe #80054-341 and stored away in the FBI archives. The Volkov case was labeled "accidental" and closed.

Mary Beth's call sparked new interest in the case for Sandy Mitchell. A prosecutor's worst nightmare is that they have missed some detail, no matter how small or seemingly insignificant, that would've changed the outcome of an investigation. And she knew Mary Beth and her skills, experience, and, most importantly, her stunningly accurate intuition as an investigator. In scanning the record on her computer screen in front of her, she noted the dead-end DNA results and continued scanning to see if any other DNA samples were taken from any other part of the boat, maybe a life jacket, beer bottles, closed containers like a cooler, or waste containers, strands of human hair, or blood samples from Volkov... anything. There were none listed on the evidence list. Sandy picked up her phone and dialed the evidence agent at the FBI lab and storage facility in Quantico, Virginia. She instructed the agent to locate the physical evidence from the sinking and provide her with a complete schedule of absolutely everything in storage, from every plank of the boat down to every strand of hair.

She received an encrypted email the next morning from Quantico:

To: Assistant United States Attorney Sandra A. Mitchell, Esq.
Re: Investigation no.: 6031-4728-0935, Volkov

Please note the hull and machinery remnants of the 1951 Chris-Craft Riviera wooden craft, serial no. 29846 were sold as salvage. Following is the evidence inventory list of items remaining in evidence, as requested.

Found within the hull of the vessel:
1. Type AA fire extinguisher
2. Ronco Flare gun set, 6 unused canisters
3. Four USCG-approved life jackets, adult size
4. Sears handheld air horn
5. waste container, empty
6. Six cans of 10W40 motor oil
7. Broken windshield glass-160 small shards

Debris found afloat or drifted ashore:
8 Fourteen empty Corona beer bottles in a black garbage bag
9. 15-gallon red and white cooler
10. Ten unopened bottles Corona beer in a red and white cooler
11. Two men's tennis shoes, size 9
12. Plastic yellow and black beach bag
13. Four yellow beach towels (in beach bag)
14. One pink "flip-flop" approx. size 4
15. Zip-lock plastic bag containing 28 grams cannabis with a small clear/green glass pipe.

Recovered from sole survivor:
16. Plastic floatation pillow
17. Torn short-sleeve tan shirt

Sandy scanned the list. At first, nothing struck her as important. But then she read the Coast Guard accident report, the description of the condition of the recovered boat, and the Coast Guard accident reconstructionist's opinion of how the accident happened. It was at high speed for a boat and violent. She took time to envision the scene.

"160 shards of glass!" she said aloud to no one. "And Volkov would've been at the steering wheel when it collided with the railroad tie! He would've been thrown through the windshield!"

She scanned the rest of the Coast Guard report and found his name: Ensign Robert Broderick, the captain of the Defender-class rescue boat that picked Volkov up from the tour boat.

Five minutes later, she had him on the phone.

"Blood?" Broderick said, "was there blood?"

"That's what I'm asking," Sandy said. "Was there blood?"

"Well, unless he was wearing a red and brown shirt, hell yes there was blood. He was drenched in it."

"The list of evidence lists a tan shirt," Sandy said.

"Might've been tan, but it was soaked in blood. That's not something you forget."

"Perfect!" she said.

"Perfect? What's perfect about that?" Broderick asked.

"Well, what I mean is, that's exactly what I needed to know."

"OK. Whatever."

"Thanks, Ensign."

"Captain," he said.

"What?"

"Captain. I said Captain. I'm Captain now."

"Oh, well, congratulations....*captain*."

"Thank you...*counselor*."

She clicked off the call without putting the receiver down and dialed the FBI agent in charge of referrals to the Federal DNA Database Unit. She ordered the tan, blood-soaked shirt be sent for DNA analysis and hit confirmations against people whose profiles are in the National DNA Index System.

Let's just see who this 'Volkov' really is, she thought to herself. Boy, I miss working with Mary Beth.

***

**Ithaca**

Detective Ken Ludlow reached Tom Foster FBI Special Agent on Foster's cellphone.

"Hey, Ken. What's up?"

"I got a passport I need you or somebody at the FBI to take a look at."

"That's easy enough," Foster said.

"Great. How do I get it to you, or to whoever?"

"Well, you must have your radar up, Ken. I'm almost to Becker's office."

"Really?"

"Yup. Got some news for him on the Weber murder."

"Good or bad?" Ken asked.

"That depends on who you're asking. Can you bring the passport down to the DA's office? I'll be there in about ten minutes."

"I'll head down now."

"See you there."

***

The three met in Gary's cramped office on the third floor of the courthouse; Gary behind his desk, and Foster and Ludlow in the uncomfortable straight-back metal chairs in front of it. Ludlow was used to the county low-budget furniture; Foster was used to the plush leather and mahogany furniture in his office at the federal building in Binghamton. He squirmed trying to get comfortable.

"Wow! All this power in one little room," Becker mused.

"Yeah," Foster said. "Too much for these humble little digs!"

"Well, we're not the federal government—unlimited budget and all," Becker said. "What brought you down here, Ken?"

"This," he said, as he handed the Volkov passport to Gary.

"So?" Becker said, setting it down on the desk, and opening it. "This have anything to do with my cases?"

"Actually no," Ken said. It's for Foster."

"Well then here," Gary said, handing it to Foster.

Foster opened it, and chuckled. "Nice smile!" he said. "There's a red flag if I ever saw one."

"That's what I thought," Ken said.

"I'll get it to our documents guy, and get back to you."

"You said you had news for me," Becker said.

"I do, and you're not going to like it."

"Well, let me have it."

"Sure. No GSR, not a speck of gunshot residue anywhere on Alina Weber. Nothing—not her hands, fingers, wrists, chest, neck, clothes—nothing. There's no way she pulled that trigger."

"Shit," Gary blurted. "Maybe she changed, showered."

"Not a chance, Gary. I already told you. Deputy Wilson was in there so fast, the gun was still hot, and the door to the house was wide open. And we didn't leave it at that. CSI searched for any sign she'd washed up and changed clothes. Nada. Nothing. She didn't have the opportunity."

Gary slumped back in his chair. "So you're telling me what? That there was another person in there that shot Weber?"

"I'm not telling you that. All I'm saying is there is no way that girl fired that gun."

"Well, suicide is out of the question," Gary said.

"Yeah. No shit," Ken said. "Three shots to the head. Usually the first one does the job."

Gary turned to Ken. "You know I have to disclose this to DeMarco."

"Yup."

"Maybe offer a plea?" Foster suggested.

"With DeMarco? Are you kidding? I don't remember the last time he was willing to plea a client out. And he'll have a field day with the GSR report."

"Vince maybe, but isn't Mike DeMarco lead counsel on the case?"

"Mike?"

"Yeah."

"He's tougher than the old man," Gary said.

"Well. You're right. Same thing with federal cases. It's gonna be a trial," Foster confirmed.

"Fuck." Gary said. "Fuck. We're back to square one. We gotta find a way to put that gun in her hand."

"Forget, it Gary. That's just not going to happen. We got a murderer out there somewhere. Maybe a co-conspirator, but a murderer out there nonetheless," Ken said.

"Ah!" Gary said. "Conspiracy to commit murder! Same as murder itself."

"What was that drug dealer's name? The one the FBI tied to the 'pound of flesh' shooting?"

"Red something," Ken said.

They all sat, thinking...

"Big Red!" Ken blurted. "Big Red."

"James Curtis, street name Big Red," Foster added.

"You thinking what I'm thinking?" Becker asked.

"Maybe. Maybe," Ken said.

"Yeah. Let's track him down," Becker said, turning to Foster. "This one's on the FBI— illegal firearm and crossing state lines."

"This is getting crazy!" Ken said. "So now the theory is this guy, Curtis, comes up here has the murder gun, goes to Weber's house, and kills the guy right in front of the guy's wife?"

A long pause...Gary's scribbling circles on the yellow legal pad in front of him on his desk.

Finally, Gary looks up. "We really need to talk to Sean Adams."

Ken nodded. "Yeah. He's the connection between Weber and this guy Curtis, 'Big Red'."

Foster gave a chuckle. "We talked about this before. You know, you can't get anything out of him. Anything between him and Dr. Weber, or even him and Curtis, is attorney-client privileged. And I understand he's even been up to see Alina Weber, so anything between them would be covered, too. You'll get nothing from him."

"There are exceptions," Gary said.

"You mean the 'crime-fraud exception'?"

"Yes," Gary said. "there's no attorney-client privilege when the services of the lawyer were sought or obtained to enable or aid anyone to commit or plan to commit what the client knew was a crime or fraud."

Foster chuckled. "Memorized that for the Bar Exam, did ya?"

"Very funny, Tom. We all had to memorize it, remember?"

"Long, long time ago, Gary. But that's a huge threshold to get over; you'd need some pretty convincing evidence to break that seal. Let me track 'Big Red' down; see what we can find out from him."

"Fair enough," Gary said.

"I'll do whatever I can to help," Ken piped in.

# CHAPTER TWENTY-THREE

Over at the Circles Office building, Attorney Sean Adams sat leaning back in his chair behind his desk as far back as the chair could go, his arms crossed tightly in front of him.

Helen Stinson Weber, Dr. Weber's ex-wife, sat in the chair in front of Evan's desk, leaning forward, glaring at Adams, her hands on the edge of his desk.

"I don't understand!" she said.

"I know; I know," Adams said, "but listen, that's just the way it goes."

"Oh, 'the way it goes,' huh? Well, that's not what you told me!"

"I was wrong."

Helen let out a loud "HA!"

"Well, I'm sorry, but—"

"But nothing! I don't believe you! My name is on that policy, I've been paying the premiums for years, I'm the goddamned named beneficiary, and you TOLD me I would collect that money!"

"Helen—"

"Don't you fucking 'Helen' me! We've got a serious problem here, Adams! And I don't trust you."

She pointed to his law degree hanging on the wall behind him. "Mr. High and Mighty lawyer! You listen to me! I paid that policy; I'm the named beneficiary on that policy! You get me that fucking money, or I'll have you disbarred! You understand?"

"I tried to tell—"

"Don't you tell me anything!"

*If she only knew Alina would get that five million—if she's not convicted of murder, that is*—Adams thought. He leaned forward in his chair, putting his elbows on the desk, his head in his hands.

With that, she stood, swiveled, and swung the door open to leave the office.

"What the fuck, lady!" The six-foot-five, red-haired, red-bearded man said as she stomped out the door, shoving him out of her way. "What the fuck, lady!"

He was wearing a black hoodie, baggy blue jeans, and Timberline boots.

He watched her as she stormed out and down the hallway. He turned, entered Adams office and closed the door behind him. Adams then heard the four words he never, ever wanted to hear in his entire life:

"I need my piece," Big Red said, extending his hand as if Adams could just put it there.

Adams sat stunned.

He looked at Big Red standing in front of his desk, and from where Adams sat, Big Red looked to be ten feet tall. Big Red put his hand down, walked toward Adams desk, flung one of the client chairs across the room, leaned over the desk, standing with both hands flat on the middle of the desk, looked Adams in the eye and said, "I need my piece. Now!"

*I'm just having a nightmare*, Adams thought. *It's just a nightmare.*

Big Red tilted his head, moved closer to Adams—almost nose-to-nose across the desk, and said, "You know who I am?"

"I do," Adams said, feeling weak with a sense of total

surrender. "I know who you are. You're James Curtis…Big Red."

"Alright, then," Curtis said, standing upright, "I need my piece. Now, hand it over."

Adams felt his eyesight begin to tunnel and took a deep breath.

"I…I don't have it here."

"Well, then go fucking get it."

Adams realized if Curtis knew the gun was with the DA or FBI, Curtis would probably kill him.

Adams attempted to gather his wits and said, "Look, I have court in fifteen minutes, and if I don't show, they'll start calling here; probably send a bailiff over to get me. I can get you the gun tonight."

"How about this?" Curtis asked, "How 'bout I wait here? You go do your court thing, then get my gun and bring it back here. Understand?"

Adams began rapid-blinking and rubbing his eyes, trying to focus his eyesight.

"What the fuck's the matter with you?" Curtis said.

Adams's eyes rolled back into his head, he blacked out, his head hitting the wooden desk like a coconut with a hollow thunk. Curtis shook his head, walked over to the coffee stand, picked up the half-full pot of hot coffee Adams had left on, came back and poured it on Adams's head. Adams awoke with a scream.

"OK! OK! OK!" Adams screamed. "I'll get you your gun!"

"Oh, hey asshole, why don't you fucking scream it to the whole world, you dumb motherfucker! Go wipe yourself, do your court thing, and get me my piece. I'll be right here waiting for you. You don't show up; I'll find you."

Adams stumbled into the small washroom and tried to clean himself as much as he could with a small hand towel. Went back to his desk, randomly picked up a file, looked at Big Red, and said. "I'll be right back."

"And I'll be right here," Curtis said with a wide grin. "And I'll be taking the fifty-grand retainer I gave you back, too. Better get that, too."

Adams stumbled and nearly tripped as he tried to get out of the office past Curtis. As he found his way around him to the door, Curtis gave him an open-handed slap across the back of his head, sending Adams falling forward to the ground in the hallway. He pulled himself up as fast as he could and ran down the hall, still stumbling as he went. He had no plan, no bright ideas, just shear panic and fear. All he knew was that he had to get away.

Curtis sat in Adams's chair, feet up on the desk and his hands behind his head with fingers entwined, with a smirk on his face. He had the opportunity for a major drug buy, and he needed that money and the gun, which he thought had never been traced.

***

FBI Special Agent Tom Foster and Detective Ken Ludlow were just coming down the front steps of the courthouse from their meeting with Gary Becker, DA when they saw Adams leaving the Circles Building in a sprint.

"Check it out," Ken said, nudging Foster and pointing to Adams, "speak of the devil!"

"He's headed to his car. I wonder what's the rush?" Foster said.

Ken called out, "Attorney Adams! Hey! Over here!"

Adams froze dead in his tracks and turned toward them.

As the two approached Adams, Foster said, "What the hell happened to him?"

"Jesus!" Ken said. "He's a mess!"

Adams's black hair was plastered to his head, his white shirt still drenched in black coffee, and he was wild-eyed.

"I gotta go!" Adams blurted. "I gotta go," as he started for his car again.

"Whoa, whoa, whoa," Ken said. "You stop right there! I said stop!"

When they got right up to Adams, they saw he was trembling uncontrollably.

"Sean, what the hell is going on?" Ken asked.

Adams couldn't bring himself to speak and looked at them with a vacant stare.

"Christ," Foster said, "he's not OK."

"Adams! Adams! What's going on?" Ken said, seriously concerned.

Adam's brain was in full-spin. He opened his mouth to try to speak, but a sudden, overwhelming sense of dread and doom filled his chest with the weight of an anchor. He began to sweat profusely, his vision went dark, and he passed out in Foster's arms. Foster slowly lowered Adams to the sidewalk, holding his head. Ken promptly called for an ambulance on his cellphone while Foster, kneeling on the sidewalk beside Adams, began checking his vital signs.

"He's out cold, and his pulse is through the roof," Foster said. He pulled Adams's tie loose, tore open his coffee and sweat-drenched shirt, loosened his belt, and elevated his head in his hands.

"I'm no doctor, but I think he's having a genuine full-blown nervous breakdown," Ken said.

"Should we transport him?" Foster asked.

"No. No. Let the first responders do that; they're trained for it."

Just then, the ambulance—a white and red van—came speeding down the street, lights flashing, siren blasting.

That got Big Red's attention. He walked over to the office window, looked out and saw Adams on the sidewalk, Ken Ludlow standing beside him, and Foster on his knees holding Adams's head. The ambulance pulled up to the curb; two EMTs rushed to Adams; the state police and others began flooding out of the courthouse to see what was going on.

"Shit!" Curtis said. "That fucker."

He quietly searched Adams's desk, file cabinet, and washroom for his gun. Finding nothing, he left the office in search of a back exit. At the end of the second-floor hallway, a staircase led down to a fire exit in the back of the building. He sprinted down the stairs, opened the door, left the building, and disappeared.

Adams drifted in and out of consciousness as he was loaded into the ambulance.

"Let's go," Ken said to Foster. "We'll take my car; I have a siren."

So, the ambulance sped, siren screaming, lights ablaze, followed by Detective Ken Ludlow and FBI Special Agent Tom Foster in the black, unmarked car—siren blaring, lights flashing, to the hospital.

***

Adams was unloaded at the hospital emergency room entrance; Ken parked his car behind the ambulance, and he and Foster began exiting, only to be stopped short by the hospital security guard.

"Hold on there! What the hell you think you're doing?" The guard asked as he quickly approached them.

Ken flashed his deputy's badge, "Police."

"I don't care if you're the president; you aren't going to park here! You'll block the next ambulance to come in. If you're a cop, you should know better!"

"Shit," Ken said, embarrassed. "No problem. Sorry."

Ken and Foster got back into the car, swung it around out of the emergency entrance area and parked in the hospital parking lot.

Adams started coming to his senses as the ambulance had pulled up to the hospital. By the time he entered the ER, he was begging to be let go and was babbling about life insurance and a Big Red.

ER Dr. John Bleecher, MD, took no time at all to triage Adams. He gave him a shot of Propofol to sedate him and had the EMTs take him straight to the psych ward on the third floor. He had enough work to do with broken arms, COVID cases, assault victims, lacerations and concussions to deal with what appeared to be a clear nervous breakdown case.

By the time Ken Ludlow and Tom Foster got to the ER department, Adams was gone. Ken went to the nurse at the admissions desk and asked where Adams was. She tapped a few letters on her computer, squinted, tapped a few more, and looked up.

"Seventy-two-hour hold," she said. "Third floor, but you can't go up there."

Ken flashed his detective's badge, then seeing the nurse roll her eyes at it, gave up, tucked it back into his pocket and turned to Foster.

"Seventy-two-hour hold," Ken said. "Nobody's going to talk to him for at least three days."

"I was afraid of that," Foster answered. "Let's get out of here."

"I better tell Becker," Ken said as the two headed out the door.

"Oh, I'm sure he knows by now."

Sean Adams had only intended to be in his office for a short meeting with Helen Weber. He'd parked his red BMW in a handicap parking space behind the courthouse. Security had already called for a tow truck.

# CHAPTER TWENTY-FOUR

## Annapolis

Randolph Everts, Esq., Victor Bennington's estate attorney—the last owner of Vince and Mary Beth's new boat, John Holdings, John Miller, the owner of the 1929 75' Elco, captain Randy Spaulding, Bill Brown, boat broker and Rick Anders, the marina mechanic met on the back deck of Bill Brown's Chris-Craft yacht. They each had a St. Louis Tommy cognac glass in their hands, and Brown had poured them each a good dose of Courvoisier.

Rick was a nervous wreck, but he understood what was going on, and he knew he'd be the one to carry it out.

"I don't see another way out," Bill Brown said. "Especially now."

Attorney Randolph Everts confirmed. "She knows too much."

"I fucking can't believe this," Holdings said.

"Oh really? You can't believe it, Holdings? You're the biggest part of this shit-show."

"Fuck you, Everts! Fuck you!"

"Hold on, hold on!" Bill Brown said, holding up a hand. "We're all in this together, and we need to fix it together."

"Some fix," Rick said.

"Well, I don't really care what you think, Rick," Holdings said. "Who would ever have imagined somebody like her would show up and start making all these connections."

"So, I take it the night cruise is out of the question," Captain Spaulding said.

"I'm afraid so. She's no Virginia Bennington, and he's no Victor," Bill said. "They'd see through it in a minute."

"It went off without a hitch with the Benningtons," Spaulding said.

"Without a hitch? Without a hitch?" Everts blurted. "Are you fucking kidding? We had to off Victor too, or did you forget about that?"

"We had no choice," Holdings piped in. "When he realized we're all 'widowers' here, he was going to the police."

"Well, we all made that choice. 'It's either me or that damn boat!' We were all given that ultimatum, and we all chose our boats," Bill said. "And I don't think any of us regret it."

Holdings shook his head.

"And I still don't think we got all the blood out of the Matthews. Have you ever noticed when the bilge pumps go off, the bilge water is pink? You ever notice that? And just why do you think that is? And how about that complete cluster fuck over on the Potomac River? Holdings killed two innocents to get the target! A Navy Cadet, for chrissakes! And now that Mary Beth, or whatever her name is, is onto me. She's been digging into Victor's estate, for fuck sake."

"Fucking FBI agent and a fucking lawyer," Holdings said.

Everts turned to Holdings. "Well, 'Mr. Volkov,' maybe it's time for you to drop that act," Everts said.

"Speaking of which, have you spoken with our buddies at Russianbryde?" Bill asked Holdings.

The mention of Russianbryde brought Holdings a wave of

nausea. He was in mortal trouble with them over the bail money for Alina Weber he'd failed to get back to them.

"Not yet," Holdings said. "I'll fill them in when we have a final plan."

"We do this with them on board their boat, *MY CHOICE*?" Rick asked.

"I haven't decided yet," Brown said. "I think it depends on how you can rig it."

"Well, I figure, an electrical short just under the port fuel tank," Rick said.

"An electrical short? You think that'll do it?" Everts asked.

"No stupid. It would have to be a small detonator disguised as an electrical problem. I'd need to make a trickle leak in the tank. Just enough to build up enough vapor to make it happen and happen fast. It wouldn't be the first old boat to blow up from a fuel leak.

"If you can pull it off that way, Rick, then we can detonate it while they're both on board."

"Not a problem, Bill. Any idea when they're coming down next?"

"I'm pretty sure they'll be down this weekend. But I'm not sure."

"Well, I'll need at least a couple hours heads-up before they get aboard."

"This'll happen out on the Chesapeake?" John Miller asked.

"Either way. Could happen at the dock. Who knows if or when they'll take that boat out on the water again," Rick said.

"Can you contain it? I mean, we can't have any damage to the marina dock or roof," Bill said.

"I can contain it to the hull and interior. Wouldn't take much. The fuel tanks are right under the cabin where they sleep," Rick said. "It'd be a hell of a blow, but that boat is built like a tank, and the explosion would probably take out the aft cabin with them in it, the back deck above, and the hardtop over that."

"How about fire?" Miller asked.

"Yeah. There'd be fire. We'd have to be ready for that."

"Doesn't that boat have an automatic fire extinguishing system?" Miller asked.

"Actually, yes," Rick said. "That might very well contain the fire to the aft cabin, but we'd still need to be plenty ready".

"Hm. Maybe a timer or remote detonation with them out on the water instead of at the dock?" Bill asked.

"I can do that, Bill."

"I like that better."

"OK. That's the way we'll do it. I can rig everything now."

"Even the fuel leak?" Miller asked.

"Oh yeah. It only has to be a pinhole. They'd hardly notice the smell, especially with the heat or air conditioning on."

"Well, we need to get them down here, and get them down here quick," Brown said, "this can't go on long, or we're all in trouble. This whole thing is going to blow wide open, and that fucking FBI agent, what's her name?"

"Mary Beth," Rick said.

"Right. That fucking FBI agent Mary Beth is the only one that can bring us all down."

"Has he ever come down without her?" Miller asked.

"No," Bill said. "Never."

"OK then, maybe you or Rick could call him, tell him there's a problem with the boat, and they better get down here quick."

"I like that," Rick said. "I like that a lot. That could work."

"Good. I'll rig what I need to and call Vince," Rick said. "I'll tell him we got a real problem with his boat, and I need him down here pronto."

"How do we get them to take the boat out on the Chesapeake?" Holdings asked.

"That's easy," Rick said. "I'll tell them I made a repair, and they need to do a quick test run. I'll make it sound perfectly safe for that."

"OK, then, gentlemen, we have a plan," Bill announced. "Make your calls, Rick."

"I'm on it, Bill."

"Alright, everyone. Have a good day. Rick, you stay. I need to talk to you."

"Sure," Rick said as the rest left Bill's boat.

They waited until the others were down the pier and out of earshot.

"It was good you told Holdings to stay away and not show up at the meeting," Rick said.

"No shit," Bill responded. "With this Mary Beth in the picture and the Russianbryde murder case up in Ithaca, the fuck up on the Potomac, the Virginia Bennington thing, Holdings has become a real liability. She would've put it all together just seeing Holdings down here."

"I agree," Rick said. "That makes me sad."

"Well, yeah, me too, but a liability is a liability, and he's the one thing that connects all that shit together. Luckily, that FBI bitch hasn't seen Holdings down here yet, But Holdings told me she saw him up in Ithaca when he went up to get Bryce's money back from the bail."

"You think she recognized him as 'Volkov' from the Potomac sinking?"

"She's fucking FBI. What do you think?" Bill said. "But she'll sure as shit recognize him if she sees him up close here."

"I think Holdings has become a dangerous liability for us," Rick said.

"See? Great minds think alike," Bill said.

"I'll run it by Bryce at Russianbryde."

"You can probably do it in person pretty soon. *EMANCIPATION* sailed from Fort Lauderdale four days ago. She'll be here soon. You ready for her?" Bill asked.

"I'm never ready for that boat, but I'll take care of her. It's getting harder and harder to find parts for those old engines. Last overhaul, I had to manufacture a header from scratch. We'll be fine," Rick assured Bill.

"We'll dock her at the end of the pier; it's the only place she'll fit."

"I figured," Rick said. "Well, I gotta go; I gotta make a call."

***

Vince and Mary Beth were enjoying a date night at their favorite restaurant, The Heights Cafe. He had his usual double bourbon on the rocks with water; she enjoyed a Gin and tonic with a touch of lemon juice. Vince's phone vibrated. He picked it up and looked at the screen.

"Hm. It's the marina," Vince said. "Nope. Not gonna answer it. They can leave a voicemail."

Mary Beth smiled, leaned across the table and gave him a soft kiss on the lips.

"Thank you, Vince. I love you."

"Cheers, sweetheart."

***

The *USS EMANCIPATION* was the U. S. presidential yacht built in 1920, purchased by the U. S. Government in 1922 and hosted presidents, politicians, movie stars, bootleggers, gangsters and their families and parties from the Roaring Twenties through the 1930s and until sold at auction by then-president Jimmy Carter in 1976. She cost about $200,000 to build and was worth millions now. In length and girth, at 100 feet long, twenty feet wide, one hundred tons—a small ship, really—she dwarfed the other yachts at the marina on the New River in Fort Lauderdale.

Now it belonged to Ron Bryce, CEO of Russianbryde, and headed up the Atlantic Coast with his captain and crew for the Chesapeake Bay to the marina in Annapolis for its annual service. Not only was Rick Anders the only marina mechanic he trusted to care for the ancient and rare 225-horsepower Winton diesels and archaic systems, but Bryce would never have to pay a dime for the hundreds of thousands of dollars of labor, parts and materials or anything else there. He owned the marina and knew every single captain's story in that marina—good and bad, and they knew it. He was in the dark, however, about the new owners of the Matthews, *MY CHOICE*.

Bryce was a "widower" from wayback.

# CHAPTER TWENTY-FIVE

## Ithaca

"I understand, Michael," Jason said. "I completely understand."

"Well, I'm sure you do, Jay, but I'm getting nowhere with her. She's completely shut down, and I don't think Alina understands what serious trouble she's in."

"Oh, she does, Mike. Believe me, she does."

"I can't get anything out of her, nothing."

"Well, look at it this way," Jason said, "supposing you were in Russia, all alone, don't know the language, no family, no friends, and you witnessed the brutal, bloody murder of your wife, and you were cuffed and thrown in jail charged with killing her; knowing nothing about anything, not knowing anything about the criminal justice system. Wouldn't you shut down?"

Michael took a minute to let this sink in.

"I hear you, Jason. But apparently, she was right there when it happened. If she didn't do it, she knows who did."

"First of all, I don't believe she did it, but isn't it possible that either she was in the bedroom when it happened? It happened fast—I mean, the front door was wide open when the police got there—and that she really didn't see what happened?"

"That's one theory, I guess."

"Or that she did see it happen and was in such shock that she's erased any memory of it?"

"Is that a thing?" Michael asked.

"Dissociative amnesia."

"What?"

"When an event is so traumatic and stressful, the brain blocks it out of the person's memory. It's not just forgetting something; it's the mind deleting it from memory. In this case, it would be localized amnesia. Localized amnesia means that someone can't recall a specific extremely traumatic event. It creates a gap in their memory."

"Like deleting a file from your computer. Erased. Gone."

"Something like that, Michael. Just could be."

"Like when a person's been subject to abuse as a child, and it only comes out years later in therapy?"

"My goodness, you're a quick learner," Jason said.

"Ha ha."

"Yeah, exactly like that."

"So, if you're right, then maybe she'd recover the memory with therapy?"

"Possible. But is there time? I mean, therapy for something as traumatic as this could take years before the memory starts to resurface."

"Maybe we file a motion for a full psychological evaluation," Michael said.

"Sure, but I clearly remember your dad telling me once that that can backfire on you because there's no way to predict what might come out."

"Yeah. He's told me that, too."

"And I doubt there's an issue of competency to stand trial. Despite the language barrier, which is pretty minor, she's very competent."

"Yup. That's not an issue."

"Well, that's what I have to answer your question, Michael."

"Thank you, Jason. As usual, you're very perceptive and very helpful."

"Good luck, Mike."

"Thanks, man, I'm gonna need it."

Just then, Mike's intercom sounded. He ended the call with Jason and hit the intercom button.

"Yes?"

"District Attorney Becker's on line one," she announced.

"I'll take it."

"I know you will," she said, hanging up the phone.

"Michael, how are you?"

"I'm good, Gary."

"Glad to hear it; glad to hear it. So look, I got some news for you."

"And?"

"That Burglary case settled. The guy took a plea."

"The one on trial?" Mike asked.

"The one that was going to go on trial in a couple weeks."

"Wasn't my case, Gary."

"I know, but since we already have a hundred and fifty people summoned in for the jury pool, the judge has moved your case up."

"You're kidding."

"Nope, but I bet you wish I was. Your Weber trial starts in ten days. Maybe you might want to call your dad. He'll be thrilled."

"You know what, Gary? Actually, he will be," Mike said.

"I know, Mike. I know."

The call ended, and Mike leaned back in his chair, arms crossed in front of him, taking time to think. He decided to hold off on calling Vince until he had at least some idea of what defense strategies might be viable. Besides, he had a dinner date with Rose this evening, and if he called his dad, there'd be no getting out of a long, long conversation or even a late-night office meeting with him. He'd call Vince in the morning. He figured nothing would happen tonight anyway. He left the office and drove to Rose's apartment.

***

When he got there, she was putting the finishing touches on an oil painting of the Ithaca Falls, with the cascade at full volume, the autumn colors ablaze. Mike was a little surprised to find her in her painting smock, a little smudge of red paint on her cheek. He gave her a quick kiss on the lips and wiped the smudge off her face. He felt oddly out of place as he sat on her futon couch in his suit and tie—a lawyer in an art studio.

"How was your day, Mike?" she asked as she swirled her paintbrush in white paint, then in blue on her pallet, and dashed it on the canvas.

To Michael, this was the challenge of this new love of his. He was used to talking to lawyers—in his practice and socially, as all his friends and prior lovers were either lawyers, cops, or detectives. This was a foray into a very foreign world for him.

"Oh, not bad," is all he could come up with.

Rose set her pallet and brush down on the small table next to her easel, turned to Michael and said, "A man of few words!"

Mike chuckled, "I had a good day, Rose. Thought about you all day."

"I'm glad," She said, "come here."

She took Mike by the hand and pulled him up off the couch. They embraced and kissed. Mike felt himself melt into her.

"So, where's dinner tonight," she asked.

"Well, I was thinking Moosewood," Mike said.

"Moosewood," She responded.

"Yes. Moosewood. It's vegetarian, vegan, organic food..."

"I know, Michael. A lovely place, but I'm in the mood for a good steak."

"A steak?" Mike asked, obviously incredulous.

"Oh, for heaven's sake, Michael! Did you honestly just assume I'm vegan because I'm an artist?"

It was the absolute truth; he did.

"Oh, of course not, Rose, not at all! But then again, I've never dated an artist."

"You going like that?" she asked, giving a little tug on his red necktie.

"Sure. Why not?"

"You need to learn to relax."

"Fine, I'll take off the tie."

"Good. Now sit and relax. I'm going to change for dinner. I'll just be a minute."

Mike sat and took time to take in her small apartment. From the looks of it, he assumed it was a one-bedroom. It was in an old four-story red brick building in downtown Ithaca that had been restored as true to the original as possible, with floor-to-ceiling twelve-pane windows and high ceilings. Other than her easel and painting supplies, the apartment was scantily decorated. The walls were lined with her works, some finished, some not.

She appeared from her bedroom wearing a black three-quarter sleeve pencil dress, black biker leather jacket and black flat mules; her lips crimson red. Michael smiled ear-to-ear at the sight.

"You look awesome!" he said.

"Why thank you, counselor! Let's go!"

They arrived at the Antlers restaurant twenty minutes later.

They agreed on a bottle of Cabernet Sauvignon. Mike ordered the lobster bisque and salmon; Rose ordered the shrimp cocktail, the slow-roasted prime rib served with au jus and horseradish, fresh vegetables, garlic mashed potatoes and green beans.

As the meal was being delivered to the table, Mike's phone sounded. He looked at the screen; it was—of course—Vince.

Mike paused.

"Aren't you going to answer that, Mike?" Rose said.

"It's my dad."

"Well then, you better answer it, don't you think?" Rose said, then turning her attention to the shrimp in front of her.

Mike silenced his phone and slid it into his jacket pocket.

"I'll call him back later."

The call stopped, then vibrated again. He reached into his pocket and clicked off the call. It vibrated again.

*Oh, for God's sake, Dad,* he thought.

"Are you OK, Mike? You look concerned."

"No, I'm fine."

"Well, is it something you care to talk about?"

Mike needed time to think; he took a spoonful of his bisque.

"I'd rather talk about the trip," he said.

Rose lit up with a smile. "The trip! Absolutely!"

"Tell me, Rose."

"Well, you know, I paint seascapes mostly, anything to do with the water. I try to capture its flow, its movement, its depth on canvas. It can be serene, peaceful or powerful, dark and deadly. So, I thought Portofino would be a perfect subject. Don't you?"

"Never been," Mike said.

"I was there once with an art class when I was at MassArt."

"MassArt?"

"Massachusetts School of Art and Design. It was so beautiful," she said, her big blue eyes widening as if she was seeing Portofino. "Will you go with me?" she asked, looking Mike in the eyes, reaching across the table and putting her hand on his.

"God, Rose. I couldn't think of anything I'd rather do, but..."

She slowly retracted her hand. "But?"

"I just found out I have a trial coming up in ten days."

"That's interesting. Can I ask what kind of case it is?"

Mike swallowed. "Murder."

"Murder? Here in Ithaca?"

"You hadn't heard?" Mike asked.

"No. I don't follow the news."

"Yeah. We're defending a young woman accused of murder."

"Oh, Michael."

"It'll be OK, Rose. I'm working with my dad on the case."

"OK! So, you win your trial, and we leave right after!"

"Your lips to God's ears, Rose," Mike said, feeling a rush of love for her and, at the same time, feeling the weight of the upcoming trial on him. He reached across the table and held her hand.

"You know Mike, you said you've never dated an artist before."

"That's true."

"Well, big surprise here: I've never dated a lawyer before."

"Oh, I'm just a guy...you know."

"What I know about lawyers is what I've seen on TV or in the movies. Is that what it's like?"

Mike chuckled, a wry smile on his face. "It depends on what you've seen. Sometimes yes, and sometimes not so much."

"Can I ask you, Michael, what is it like to be a lawyer?"

Mike took a sip of his wine and took a moment to think.

"Well, a lawyer's job is to represent his or her client—"

She cut him off. "No no, Michael, for you—what is it like for *you personally* to be a lawyer. How do you feel about it?"

"That's a tough one," Michael said.

"Really? I thought words were a lawyer's craft."

He took a sip of his wine, savoring the moment to collect his thoughts

"Alright. How do I feel about being a lawyer? First word that pops into my head is… 'stressed.'"

"That makes me sad, Michael, but I'm glad you told me. What else?"

"Worried. Exhausted," he confessed, a weariness in his voice.

"Ugh, Mike. Always?" She asked, concerned.

"I don't know, but certainly right now," he admitted, his thoughts turning to the looming trial.

"Right. You have the trial coming up soon. Is that it?"

"It is."

"It must be a heavy burden when someone's future is at stake, and they're relying on you. I can't even imagine," Rose empathized.

Mike paused, struggling to find the right words. "Well, I—"

Rose pressed further, asking the question that had cut straight through his defenses. "Are you scared, Michael?"

"Scared?" Michael repeated, with a slight smirk on his face. "Scared?"

"Scared, Michael," Rose persisted. "Are you afraid? Are you frightened?"

The question struck him like a ton of bricks. His smirk vanished, replaced by a vulnerability he had hidden away since childhood. His gaze met Rose's, and he felt an inexplicable connection, an urge to open up to her like he had never done before.

"Scared? Well, you know, of course not I…I…"

He stopped dead in the middle of the sentence and swallowed hard, then took a long sip of his wine.

She had cut to his soul. She had pierced his armor. He hadn't been asked this question since he was a little boy, a very little boy. And probably asked by his dad, Vince. He met her big, soft, blue eyes and knew he had found someone he could open his heart to let his guard down with. She was no lawyer, no cop, no detective; she had the soul of an artist, and this was all new to Michael. All his former mates had been in the law, all wearing the same armor—all the time ready to make a point, win an argument. God forbid admit any feelings, especially feelings of vulnerability, weakness, fear.

Still holding her gentle gaze, he reached across the table for her hand.

"I'm scared to death, Rose."

She took his hand in both hers and gave a small closed-lip smile, and a single tear trickled down her cheek.

With a deep breath, he reached across the table, cupped her face in his hand and gently wiped the tear from her cheek with his thumb. "Oh, Rose," he whispered.

# CHAPTER TWENTY-SIX

District Attorney Gary Becker sat across the conference room table from FBI Special Agent Alfred Platford, the FBI blood-spatter analyst. Platford was a charter member of the Scientific Working Group on Bloodstain Pattern Analysis (SWGSTAIN), made up of experts from North America, Europe, New Zealand, and Australia. He was among the best of the best.

"What I need to know is where the blood came from, how the perp was positioned, what movements the shooter made after the shooting, and whether there could have been more than one other person in the house besides the shooter and the victim," Becker said.

"I know, Gary."

Gary nodded in rapt attention as Platford launched into his explanation.

"You see, Gary," Platford began, his tone a blend of authority and courtroom charm, learned from hundreds of times taking the stand in murder trials, "in this case, it all boils down to gunshot splatter," he said, as he lifted his brown leather briefcase off the floor, set it on the conference table and opened it. He pulled a packet of eight-by-eleven-inch color photos out of the briefcase and spread them out across the top of the table.

"So what we look at is forward spatter and back spatter. Forward spatter is the blood that spatters from the exit wound, and back spatter comes from the entrance wound. All that depends on the caliber of the gun, where the victim is shot, whether the bullet lodges in the body or not, how far the shooter was from the victim and where the victim was relative to the floor, walls, and other objects. Typically, forward spatter is like a mist, and back spatter is larger droplets—not really large, but larger than forward spatter."

The first photo Platford showed Gary was of Weber's body slumped over on the couch.

"OK, so this photo was taken from exactly two feet from the victim."

Gary picked up the picture, slid his glasses down to the tip of his nose and examined it closely. "Jesus, what a mess." He said, already thinking about how the jury would react to seeing such a graphic, bloody, horrible crime scene.

"Well, yeah," Platford said. "Look here," he said, pointing to the back of the blood-soaked couch, "Right there. See it?"

"I see one bullet hole," Gary said.

"Correct. Only one bullet passed through the victim. Two were lodged in the occipital bone—the lower edge of the occipital bone. In other words, the very bottom of the back of the skull."

"OK, so?"

"So wait. Now, look at this photo. It's the wall to the left of the front door and the front door in the open position."

"Yeah. When Deputy Wilson showed up, the front door was wide open."

"Right. So look at this," Platford said, running the tip of his pencil in an arch on the photo from the floor, up, over, and back to the floor.

"OK, I see that, what's that mean?"

"That's back spatter."

"OK, I see that, but what's it prove."

Platford gave a chuckle. "Really, Becker? Think about it. Why would back spatter make an arch pattern?"

Gary took the photo and twisted it to the left and right, "I give up."

"Because someone was standing there and took the brunt of the back spatter on himself."

"Or herself," Gary said.

"Sure, or herself. Now look at this one."

Platford selected a computer-generated graphic from the pile and set it in front of Gary.

"First thing, it's a small house, and the room the victim was found in was the combination living-dining room. The entrance to the house from the front door goes directly into the living room area, and the couch the victim was found in was against the wall across from and facing the front door."

"OK, I know that," Gary said.

"Right. So, based on the one bullet hole in the back of the couch and the two bullets lodged in the occipital bone—the lower edge of the occipital bone, we know the angle of the three shots. First, we can tell they were all fired from the same distance and angle and were fired in rapid sequence. We know they were fired in rapid sequence because the body of the victim didn't move much from one shot to the next. They're close in a cluster. There was no pause between shots at all."

"OK, and?"

"So here's the rub," Platford said, smiling, as he pulled the next photo out of the pack. It was a photo of the floor between the front door and the couch. "This is what we found when we sprayed the carpet with Luminol."

Gary picked up the photo and brought it close to his face. "These lines are distance markers, I presume," he said.

"You presume correctly," Platford confirmed. "Do you see them?"

"These?" Gary asked, pointing to two spots that were not fluorescing with the Luminol.

"Yes. And note the distance to the victim."

"OK, sixty-four inches."

"Right. Sixty-four inches from the Victim's feet on the floor at the couch to the two spots not reactive to the Luminol. So what's that tell you?"

"What is this, a quiz?" Gary said.

"Sure. Why not? What's it tell you?"

Gary took his time. He looked at that photo, then the others and the computer-generated graphic, trying to put the pieces together.

The conference intercom sounded.

"Detective Ludlow's on his way back, Gary." The legal intern said.

"That's fine," Gary responded.

"Hello, gentlemen," Ken said as he entered the conference room and took a seat to Gary's right. "Ah! Good! Let me see those."

Gary handed Ken the pack of photos. Ken scanned through them in two minutes.

"From the angle of the shots, the area on the wall not showing blood spatter, and the two footprints on the carpet that showed up after spraying the Luminol, it looks like the shooter was almost on top of the victim, or at least the gun was. Gotta love that Luminol; can detect blood that's been diluted up to 10,000 times; comes in contact with blood, it lights up the whole place in blue. I'm a big fan."

"There's more, Ken," Platford said.

"What?" Gary asked, being more than a little annoyed at Ken.

"The distance from the two footprints to where the gun must've been when it was fired."

"Yeah?" Gary asked.

"Had to be a very long arm. If the shooter was standing here," Platford pointed to the two footprints, "and the angle of the shot was this," he said, pulling out the graphic, "then the shooter had a long reach. Had to be tall, maybe at least six feet, maybe more."

Gary took off his glasses, set them on the conference table, leaned back in his chair and rubbed his eyes; a headache coming on.

"Oh, DeMarco's gonna love this!" Ken said, laughing. "He's gonna fucking love it!"

"And by the way, there was a lot back blood spatter—a lot. Did you find blood spatter on Alina Weber?"

Gary and Ken exchanged glances.

"No," Gary said. "No."

"Well, OK. Look, here's my full report," Platford said, pulling a manila folder from his briefcase and handing it to Gary Becker.

He took the folder and set it on the table without looking at it and turned to Ken.

"How tall you think Alina Weber is, Ken?"

"Hm, I don't know exactly, but I'd guess probably at least five-foot-three, maybe four."

Gary turned to Platford. "What do you think?"

"I think you got a problem," Platford said.

***

"He's coming to," Dr. VanAldon said. "Maybe get him something to drink."

"Yes, doctor," the student nurse said. "I'll be right back."

"Well, good morning, Mr. Adams," the doctor said, gently tapping him on the shoulder.

Adams jolted and sat up in the hospital bed.

"How, how long, how long have I—"

"Since yesterday. We had to sedate you. How are you feeling now?"

"I don't honestly know. Groggy, I guess."

"Well, that's normal."

Adams suddenly realized he was wearing a hospital gown. He looked around the room. "Where are my clothes?"

"Right there in that plastic bag," the doctor said, pointing under a chair by the wall, "but you're probably going to want to call someone to bring you some clean ones."

"I don't have time for that. I need to get out of here."

"Not so fast. You're on a seventy-two-hour hold, and you're not going anywhere until you undergo an evaluation."

"Evaluation? I'm fine! I need to go."

"You just wait here. The nurse is bringing you some orange juice. You'll be out of here soon enough."

Adams fell back onto his pillow in despair. The doctor turned, grabbed the plastic bag from under the chair and left the room, closing the door behind him. A minute later, the student nurse returned with a paper cup of orange juice, handed it to him, tucked him in, wished him well, and left, closing the door behind her.

He waited for her to be out of the room and out of sight and slid out of the bed. He reached under the chair for his clothes and found only his phone and wallet; the plastic bag was gone. He snatched up his phone and wallet and cautiously approached the door, half expecting to be locked in. Adams peered through the narrow window in the door and waited until no one was in sight. Wearing only his hospital gown and yellow rubber-bottomed hospital-issued socks, he slowly and casually—so as not to raise suspicion—opened the door and made his way to the nearest exit. As he strolled, he tapped his Uber app and summoned his ride.

He had decided to come clean once and for all; confess completely, tell the cops Alina had nothing to do with the murder, explain where the gun came from, that it was all his doing and that she had no idea what he was going to do. He'd had it with all of this; prison would be a better life than what he was living now. He felt a sudden freedom, a tremendous weight lifting off his chest, and he felt himself smile for the first time in months.

Adams eventually made it to the front entrance of the hospital and stood by one of the square pillars to keep out of sight as much as possible. A black Land Rover SUV pulled up to the entrance. Adams scampered up to the SUV, tapped on the passenger side window, and the driver lowered the window two inches.

"Uber?" Adams asked.

"Yup! Hop right in," the driver said.

Relieved, Adams opened the back door and flung himself into the back seat.

The SUV pulled away from the curb.

"Fuck man, you must be cold," the driver said.

Adams felt a chill go down his spine; he was pretty sure he recognized that grave voice.

The SUV sped out of the parking lot and, once on the road, took off at a hundred miles per hour with Adams pinned against the back of the seat and petrified. He grasped his phone and tried to turn it on; it slid out of his trembling hands and under the front seat.

"W-w-where-where are we going?" Adams asked, struggling to speak.

"That's up to you, counselor," Big Red responded.

"Me?"

"Yeah, you. You tell me where my gun and money are, and that's where we'll go. Otherwise, I have other plans. So, it's up to you."

Adams didn't care anymore; he didn't care about anything, and he knew his situation was now hopeless.

"I'll tell you where your gun is!" he exclaimed. "I'll fucking tell you!"

Big Red craned his neck and looked back at Adams, realizing maybe, just maybe, Adams had lost his mind.

"The DA has it, or maybe the FBI, or maybe the sheriff!"

"WHAT!?" Red screamed. "WHAT DID YOU SAY?"

At that, Adams started to laugh.

"What the fuck you laughing about?"

Adams was now laughing so hard and uncontrollably that he couldn't talk.

Red slammed on the brakes, fishtailing the car until he was able to pull over to the side of the road. He opened the driver's door, got out, opened the back door, pulled Adams out by the throat, and pinned him against the side of the car, lifting him three inches off the ground.

"I killed a man with your gun! I fucking killed a man with your fucking gun!" Adams said almost triumphantly. "The cops got it!"

Big Red's eyes bugged, a boiling rage rising up in his mind and body, and he squeezed Adam's throat. Adams gasped, started to gurgle, and went into spasms, kicking wildly. Red continued until he felt a snap in Adam's neck, and he went limp; he dropped him to the ground, and Adams lay there motionless in nothing but his hospital gown and yellow rubber-soled hospital-issued socks.

The car sped off.

A cold, drizzly rain started to fall.

# CHAPTER TWENTY-SEVEN

Vince and Mary Beth finished their meal of a shared Margherita pizza, followed by Crème Brûlée and coffee and headed home. Vince's phone sounded on the car Bluetooth. He looked at Mary Beth. She smiled.

"Answer it, Vince."

He tapped the screen and said, "Hello."

"Hello, Vince?"

"Yeah."

"Hey, Vince, it's Rick. Rick at the marina. You got a minute?"

"Sure, Rick. What's up?"

"Well, I hate to be the bearer of bad news. Not real bad news, but, well—"

"What is it, Rick?" Mary Beth jumped in.

"Oh! Hi, Mary!" Rick said, surprised she was on the call.

"Mary Beth," she said.

"Right, right. Mary Beth," Rick said, now liking her even less.

"Well, found a problem with your boat."

"Yes?" Vince asked, feeling his stomach tighten a bit.

"Yeah, well, I was walking down the pier, you know, always checking things out, and I saw diesel in the water around your boat."

"That doesn't sound good," Vince said, fully aware of the possibility of a violation of environmental laws.

"Not to worry," Rick said. "I managed to get on board, check things out and found a loose clamp on one of the fuel lines that let some fuel into the bilge. Then, when the bilge pump went off, it pumped that diesel along with bilge water out into the marina."

"Did you report it to the EPA down there?" Vince asked.

"Ha! You kidding? We don't report stuff like that at this marina."

Vince looked at Mary Beth; she shook her head.

"Anyway, Vince, I think you need to get down here; take a look at everything and make sure we don't have any more problems," Rick said.

"You have it under control?" Mary Beth asked.

Rick tightened his lips. "Oh sure, but you gotta come down and sign off the repair," Rick insisted.

"I'll call you back, Rick," Vince said, "and thank you for taking care of it."

"You bet. Talk to you later," Rick said, then clicking off the call.

They drove in silence for a few miles, and then Mary Beth spoke up.

"Vince,"

"I know, Mary Beth, I know. Not good."

"You're an officer of the court, Vince."

"I know, I know. So, if there was a diesel leak into the water, we need to report it."

"Maybe. It depends on how much fuel leaked."

"I feel a little sick," Vince said.

Mary Beth chucked. "Tough guy!"

"I'm going to call Rick back," Vince said, tapping the "recent" tab on the screen. Rick answered on the first ring.

"You headed down?" Rick asked.

"What? No. But I have a couple questions," Vince said.

"Yeah?

"Did you shut off the bilge pumps so they don't go off again and pump whatever diesel is left in the bilge into the water?"

Rick took time to think.

"Well, did you?" Vince asked again.

"Oh, no. Can't do that; that's up to the captain. I can't shut off systems. Only you can do that."

Vince's chest tightened. "Are you kidding?"

"Nope. Only you can shut off systems on your boat. Only the captain."

"Well, that's ridiculous. So how about this: I authorize you to shut off the bilge pumps until I get down there."

"Nope. I can't do that. What if she springs a leak and starts to sink, and I'd shut them bilge pumps off? Uh-uh. I ain't doing that. You gotta do it. And I gotta go now. You better get down here."

The call went dead.

"Let's get home, Vince. We can go down in the morning if we leave early," Mary Beth said, patting Vince's arm. "Call him back."

Vince called Rick back and told him they'd be there late morning.

Rick was actually calling from inside The *MY CHOICE*, Vince's and Mary Beth's boat. He had already shut off the bilge pump systems and had just finished attaching the remote ignition device to the side of the fuel tank. The tiny pinhole he'd made in the fuel tank started a slow, almost imperceptive drip of fuel into the bilge. He wiped his hands on the dirty orange rag tucked in his belt, smiled, and left the boat.

***

Vince and Mary Beth got home to their lake cottage minutes later.

"I better call Michael and tell him I'll be gone tomorrow," Vince said.

"Good idea," Mary Beth said. "I'm going to get ready for bed. Don't be long."

"Believe me, I won't," Vince said, tossing her a warm smile.

Mary Beth disappeared into the bedroom, and Vince sat at the dinette and called Mike.

"Hey, Michael. I have something to tell you," Vince started.

"Wait," Mike said, "I was going to call you in the morning, but I might as well tell you now."

"Sounds serious," Vince said.

"How about this: trial in ten days."

"What trial?" Vince asked.

"Alina Weber."

"What?!"

"Yup. That burglary trial everybody was sure wouldn't settle, settled. We're moved up."

"We're not ready," Vince said, feeling a little dizzy.

"Doesn't matter; District Attorney Becker is."

"I wonder," Vince said. "I wonder."

"Well, we've got a lot of work to do, Dad."

"We sure do. Have you gotten anywhere with Alina yet?"

"No, but Jason's given me a lot to think about."

"Alright. I'll be in the office in the morning."

"Are you coming, love?" Mary Beth called from the bedroom.

The sound of her voice calmed him.

"I'm coming, sweetheart."

The bedroom was lit by gold moonlight through the window. She lay on her side on top of the bedspread, her head propped up on her hand, her long legs crossed at the ankle, one knee raised. She wore nothing but a sheer black spaghetti-strap babydoll nighty-just the top. Vince stripped down to his boxer shorts, shut out the world, and melted into her arms.

There'd be no trip to the boat in the morning.

# CHAPTER TWENTY-EIGHT

## Annapolis

When John Holdings fired up the starboard engine of his beloved 1930 Frank Anderson Yacht, it puffed a small black puff and settled into a smooth rhythm. He smiled. When he fired up the port engine, it blew a cloud of white exhaust that engulfed the boat, and the engine sounded like someone shaking a coffee can filled with rocks.

"Fuck!" he exclaimed aloud to no one. "Fuck fuck fuck!" he yelled, shutting down the port engine, then the starboard. He walked away from the helm on the back deck, opened the hatch to the stairs, scurried down the six steps to the main cabin, and opened the port engine hatch. He was immediately engulfed in dense, oil-laden, sooty smoke, sending him into a coughing fit. He slammed the engine hatch closed and rushed back onto the back deck. He took a minute to calm himself and collect his thoughts. One thing he knew for sure was that coffee can sound was about as expensive a sound as anything he'd ever heard before. He was pretty sure that the 1930 rare, irreplaceable engine had just self-consumed itself. The engine rebuild he'd done years ago had taken two years and cost over $100,000. He suddenly felt a wave of nausea and fought not to vomit on the deck.

***

Rick and Bill Brown were each enjoying a bottle of Heineken beer on Bill's Chris-Craft's back deck when Bill spotted the plume of white smoke billowing from the next pier.

"Well, shit," Bill said, "look at that."

Rick turned his head. "Oh shit, man. That's a blown engine. Where's it coming from?"

Just then, Rick's phone sounded.

Rick pulled his phone out of his pocket. "Hm. It's Holdings," Rick said.

"Speak of the devil," Bill said, taking a swig of his beer.

Rick clicked on the call. "Hey John, what's up?"

Holdings took a deep breath, trying to keep his cool. "I think I got a big problem, Rick. I fired up my port engine, and I think she may have blown."

"Oh yeah?"

"Well, she blew a shitload of white smoke and rattled like a son-of-a-bitch."

"White smoke, huh?" Rick said, feigning ignorance.

"White. Can you take a look at it?"

Bill's and Rick's eyes met. "Sure. Sure, I can take a look at it. I'll be right over. Don't do anything until I get there."

"OK, thanks, Rick. Much appreciated."

"You got this?" Bill asked Rick.

"Oh, I got this, alright."

The white smoke was hanging over Holding's boat and dock like a dense sea fog when Rick got there. Holdings was standing on the side deck.

"Permission to come aboard?" Rick said facetiously.

"Come on up," Holdings responded, still pale and shaky from the experience.

Rick went up the boarding stairs, onto the side deck, then followed Holdings onto the back deck and down into the main cabin. The boat reeked of charred rubber, burned grease, and scorched oil. It took a second for Rick to catch his breath.

"OK," Rick said. "Here's what we're going to do. Wait. First, is the ignition shut off?"

"Yeah. I shut it off."

"OK, good. What I need you to do is open the engine hatch. But wait. How long did you run it?"

"Oh God, for a few seconds, that's all. It went bad as soon as I started it up."

"That's good. Then the engine won't be hot."

"Right."

"So, I want you to open the hatch and let whatever smoke is still in there out. We'll look everything over, then we'll take it from there. Go ahead and open it up."

Holdings bent over and struggled to lift the near-two-hundred-pound engine hatch open and set it open on its hinges. The last bit of acrid smoke followed the opening hatch, and they both stood back until it mostly cleared. They stared at the old engine, covered in oil.

"OK. So here's what I want you to do," Rick started. "I need you to get down on the port side of the engine and see if the main head gasket is blown."

"How do I do that?" Holdings asked.

"Easy. Here, take my flashlight," Rick said, pulling a small penlight from his belt. "Get down between the hull and the engine and look back up to the top third of the engine. You should see the source of the oil. That'll tell us a lot."

"Alright, if you say so," Holdings said, taking the flashlight from Rick, slowly, reluctantly crawling down into the bilge and squeezing himself in the tight space between the blown engine and the side of the hull.

"You see anything?" Rick asked.

"I don't really see anything yet," Holdings said, scanning the engine with the small light.

Suddenly, the heavy engine hatch tilted and slammed shut.

"RICK! RICK!" Holdings shouted, "OPEN THE HATCH! OPEN THE HATCH!"

There was no response.

Now, in a full-blown panic, Holdings screamed, "FOR GOD'S SAKE, RICK! RICK! OPEN THE FUCKING HATCH!"

From his position jammed between the engine and the side of the hull, with the hatch shut, he was completely helpless, with no way to free himself or to reach the hatch to even attempt to lift it. Not that it would've made a difference; Rick Anders was standing on the hatch.

Then, the engine started to crank over.

Holdings let out the unmitigated, out-of-control, high-pitched scream of the damned, knowing he was about to die.

The engine fired, making a thunderous, deafening, overwhelming, high-pitched scraping and clanking sound, with boiling engine oil and diesel spewing into the engine compartment along with toxic exhaust and searing gasses.

It took only four minutes for John Holdings, aka Nikolai Volkov, to perish in that pitch-black, steaming, airless, toxic furnace.

*That was quite the opportunity,* Rick thought, as he shut down the port engine, dusted his hands off on his pant-legs, and walked off the boat. Job done. *That was just too easy.*

Ten minutes later, Rick rejoined Bill on Bill's back deck. They sat with their beers in hand.

"Well," Bill started, "Overall, he was a good guy."

"Yeah. I'd say so," Rick said. "Too bad he was such a fuck-up. We really didn't have any other options."

"I wonder what's going to happen to his 'Crab Shack'?"

"Whatever *you* want to happen to it, I'd say," Rick said.

"Hm. Maybe put it up for sale along with his boat. Should get a pretty penny for both."

"Fuck yeah. I've only ever seen another Frank Anderson yacht, and that was on Yachtworld USA, and it went for a million bucks."

"No shit!" Bill said a wide grin across his face. "But what about the blown engine?"

Rick chuckled. "The engine's fine, just blown head gasket. He shut it off before any damage was done. That's a one-hour job."

"Ok then. We'll set up the 'estate.' I'll call Everts."

"Ah right," Rick said, "attorney Randolph Everts. How is Randy these days?"

"Probably drunk and in trouble, but he handled the Victor Bennington estate so we could sell the Matthews."

"Right," Rick said. "And we'll need an 'estate' set up to sell Holdings' boat and the Blue Crab Shack."

"Exactly," Bill said. "and speaking of Holdings, you better go get him out of the bilge before..."

"I know. We don't want the same mess we ended up with in the Matthews. You can never really get rid of that smell."

"Or the pink bilge water."

"You mean *blood* bilge water," Rick said.

Bill glared at Rick.

"And I want to get top dollar for that Frank Anderson yacht."

"Right," Rick confirmed.

"Eh. We should've run this by Bryce first," Bill lamented.

"What, and miss this perfect opportunity? It was handed to us, for fuck sake. He won't care."

No one would come looking for John Holdings; his wife had "passed away" a year after he bought his beloved 1930 Frank Anderson Yacht ten years earlier—another "widower" in the marina. They'd had no children.

***

Meanwhile, the former presidential yacht, *EMANCIPATION*, with Ron Bryce standing at the helm next to the ship's captain, powered north past the Thomas Point Shoal Lighthouse at twelve knots, just hours away from Annapolis. Bryce's eyes stared forward, focused on nothing. The closer they got to Annapolis, the hotter his blood boiled. The place that was to be his pride and joy and respite had become his worst nightmare thanks to a bunch of imbeciles running the place. He knew it would soon all come crashing down on him, and he would probably end up spending the rest of his life in prison—or worse.

# CHAPTER TWENTY-NINE

It was seven o'clock in the evening, and District Attorney Gary Becker was still busy organizing all the evidence, reports, photos, and affidavits he was required to disclose to Michael DeMarco before trial. He could hold nothing back, no matter how devastating it might be to the prosecution. The law in New York had changed in 2020, requiring the DA to voluntarily disclose all his evidence to the defense counsel on an "accelerated timeline." And time was running out.

His intercom sounded.

"Gary, attorney Mitchell is on line three for you."

"Mitchell?"

"Yeah. Attorney Mitchell."

"Take a message. I honestly don't have time right now."

"Gary, she's calling from the Department of Justice."

"Why didn't you say so?"

"I just did."

"Alright, put her through."

Gary punched the line-three button on the desk phone. "This is Gary Becker."

"Well, hi Gary, how have you been?"

"Uh..."

"This is Sandy Mitchell, Assistant U. S. Attorney. We met at a conference down here at the United States District Court for the District of Columbia, Washington, D. C."

Gary sat up straight in his chair.

"Ah, yes. I remember. The RICO conference."

"Exactly."

"And how could I ever forget the title of your presentation: 'RICO-Organized Crime's Worst Nightmare,'" Gary said as the memory of the six-hour-long, droning-on lecture came painfully flooding back.

"That was it! How kind of you to remember."

"So, what can I do for you, Ms. Mitchell?"

"Call me Sandy."

"Ok, Sandy. What can I do for you?"

"Well, seems you've got a murder case about to go to trial in a few days."

"I do."

"We down here have an interest in that case."

"How so?"

"Our intelligence tells us the case you're about to try may be the tip of the iceberg of a continuing criminal enterprise."

This gave Gary mixed emotions. He was pretty sure he knew what was coming next; that she was about to tell him the Feds were going to take over the Weber murder case as part of a CCE, continuing criminal enterprise case. That was bad news since he looked forward to the first murder case he'd tried in years. On the other hand, with the materials he was about to hand over to DeMarco, there was a damn good chance there'd be an acquittal or at least a hung jury, and that would be bad for his record as a prosecutor. So maybe this was good news.

"We're taking over the Weber murder case," Sandy said, getting right to the point.

"Since when is murder a federal case?" Gary said.

"Well, let's put it this way, Gary. Murder aboard ships in the United States territorial waters come under federal jurisdiction."

"This murder wasn't on a ship, for chrissakes!"

"No, but our intelligence implies she's part of a criminal enterprise—a conspiracy—that has committed murders onboard vessels in national waters."

"Murders? More than one?" Gary said, astonished.

"18 U.S.C. Section 2280, and Federal Rule 24, Intervention by a Government Officer or Agency. 'On timely motion, the court may permit a federal or state governmental officer or agency to intervene if a party's claim or defense is based on a statute,' and that statute is 18 U.S.C. Section 2280. Look it up, Gary. That's all I can tell you," Sandy said. "We filed our motion on an emergency basis this morning, and you and defense counsel will be served with it this afternoon."

"Well, your timing is bad, Sandy. The jury pool has already been subpoenaed and is in place, and we're ready to go. The judge isn't going to..."

"Are you going to oppose it?"

"What?"

"I said, are you going to oppose my motion to intervene and take over the case?"

"I don't know..."

"Well, your trial is off the calendar. This I know. The judge adjourned it as soon as we filed our motion."

Gary leaned back in his chair, breathing a sigh of relief.

"Alright. I don't think the state will oppose your motion."

"OK, good, and Gary..."

"Yeah?"

"Have you done your disclosure to the defense counsel yet?"

"No. I was just about to."

"OK. Don't. And I'll be needing your entire file. Do I have to subpoena it?"

"No no. I'll gladly hand it all over to you."

"And have detective Ken Ludlow contact me as soon as possible."

"Ken? Sure. But Sandy, listen..."

"Yeah?"

"I'm convinced that if Alina Weber didn't do the shooting, she either had someone do it for her or was, at minimum, a co-conspirator."

"Oh, really?" Sandy asked.

"Yeah, really," Gary said. "I'd be taking a real close look at the murder weapon, its history, and a guy named James Curtis, street name Big Red."

"Alright, I'll do that, Gary."

"I appreciate it, Sandy."

"Good. We'll stay in touch."

"Just let me know what I can do to help."

"Thanks, Gary."

"You bet."

***

Michael received his first and only notice of the cancellation of the trial by NYSCEF, the New York State Courts Electronic Filing system. The notice appeared in his email on his desktop computer. He clicked on it and signed in. It was a simple, straight-forward letter from the judge:

"Counselors:

The trial in the matter of *The People of the State of New York v. Alina Weber* Docket no.: 2023-445321 has been adjourned until further notice. The jury pool has been discharged. There will be no further action on this matter in this Court until further notice.

Hon. Jennifer Britton
County Court Judge
County of Tompkins"

Michael was stunned. He read the letter over three times. The first thought that came to mind was that Alina had retained new counsel and had taken a plea deal behind his back. He felt queasy. He picked up the phone and dialed Gary.

"Did you call off the trial, Gary?" Michael asked.

"Nope."

"Did she retain new counsel?"

"You mean, did she fire you?" Gary said.

"Well, yeah..."

"Not that I know of."

"Did she take a plea?"

"Good grief, Mike; you would know if she did. I don't work behind people's back."

"Well, what the hell, then, Gary?"

"Feds."

"Feds?"

"Yeah. I got a call from the DOJ. They're taking over the case. The U. S. Attorney filed a motion to intervene this morning. You can expect to be served any minute."

"What? This is no federal case!"

"Mike, you're not going to believe the basis for the motion."

"Yeah?"

"Yeah. 18 U.S.C. Section 2280."

"Wait. That's the pirate statute! What the hell!"

"You can call it that, but it applies to murders on vessels on national waterways."

Mike leaned back in his chair to the stop, his head spinning. "I don't..."

"I didn't understand either, Mike. What I got from her so far is that your client is a part of some conspiracy, and there were murders in that conspiracy at sea."

"Fuck. And anyone in a conspiracy is guilty of all the crimes committed in that conspiracy even if they didn't have anything to do with it."

"You're sharp, kid, what can I say."

"Are you going to oppose the motion to keep the case here?" Mike asked.

"Nope. She can have it."

"So now?"

"It's going to the feds. You and Vince are both members of the federal bar, aren't you? You can defend her in federal court."

"Yes. But I don't understand. What else did they tell you?"

"Better sit down, Mike."

"I'm sitting."

"The U. S. Attorney says this case is part of a Continuing Criminal Enterprise, a conspiracy."

Mike sat speechless.

"Mike? You still there?" Gary prodded.

"Yeah, I'm here. So they say Alina is part of a conspiracy? CCE?"

"I don't know. Maybe a RICO case."

Mike let out a chuckle. "Racketeer Influenced and Corrupt Organizations Act? You must be kidding."

"I'm not kidding, and believe me, they're not kidding either. Could be RICO; that would be the Fed's thing."

"This is insane," Mike said, "insane."

"And I should probably tell you, I'm handing all my disclosure over to them."

"You mean the disclosure you owe *me*!"

"Not anymore," Gary said. "Maybe you should call the U. S. Attorney on the case. Her name is Sandy Mitchell, out of the Washington D. C. Branch."

Mike thought he'd heard it all until this.

"D. C.?"

"Yeah. I said, D. C. And good luck."

They ended the call, and Mike sat motionless in his chair, staring at the judge's letter still up on his computer screen. His mind spun:

*Was Alina a part of a conspiracy? Did she kill Dr.Weber in furtherance of organized crime? Is she that good of an actress that she's fooled all of us? What to do next? Call Vince? Get a hold of Alina? Call the DOJ? Is this over the five-million-dollar life insurance? Are we all just fools? Can I even still represent her? No way I can if she's been concealing things and lying to me!*

Suddenly, a wave of panic flowed over Michael. He pounded his fist on his desk.

"We have no discovery! No disclosure!" he said aloud. "Nothing! We are totally in the dark! No wonder we can't get a thing out of Alina!"

Just then, the door swung open, and Miranda popped into his office.

"Michael, are you alright? You're talking to yourself."

"No, Miranda. I'm not."

"I'll get you a water," she said as she turned on her heels and left his office. She'd never seen him like this before.

A minute later, she came back into his office, handed him an ice-cold bottle of Fiji water, and plunked herself down in one of the client chairs in front of Mike's desk. She sat in silence, a soft gaze on Michael.

Michael opened the bottle, took a long, refreshing sip, and turned to her.

"Something very wrong is going on with the Weber case," he said flatly.

"Wrong? Did we do something wrong?" she asked.

"I don't think so, other than maybe to be duped by the client."

"Alina? No way."

"I don't know. I really don't know. All I know right now is that the Feds are taking over the case, and our trial is canceled."

"The trial's off? Well, that's good news, isn't it?"

"Usually, I'd say so, but apparently, the Federal Government thinks she might be part of a conspiracy."

Miranda was completely baffled.

"Well, this is beyond my pay scale," she said. "Whatever it is, Mike, you can handle it. I know you can."

"Thank you, Miranda."

"You'll see, Michael; you're going to kick ass like you always do, no matter what."

"Your lips to God's ears, Miranda, your lips to God's ears."

"Look, Michael, it's late, I'm tired, and I'm gettin' out of here. Why don't you take a break and go have a nice dinner with that cute artist you're dating. Maybe talk about something other than your cases for a change."

"That sounds great, Miranda. You have a good night," Mike said while knowing how impossible it would be to sit across Rose for dinner and not be consumed by what was going on in the Weber case.

"Oh, and time-and-a-half, remember?" Miranda said.

"Of course," Michael said, "of course."

Mike waited until he heard Miranda leave, pulled up the website for the D. C. Branch of the United States Attorney's office and dialed...he got voicemail. He ended the call and dialed Rose's number. She answered on the first ring.

"Dinner tonight?" he asked.

"Love to," she said.

***

Bob Mix was driving home through pounding rain in his Subaru after a long day at work at Cornell University, where he was security supervisor at the North Campus when his headlights briefly illuminated something laying on the shoulder of the road. It was dark and raining, and although he couldn't make it out, his training and his natural instincts caused him to slow the car, pull over onto the side of the road, put his flashers on, and slowly back down the road closer to the object. He got out of the car and approached the heap carefully, wishing he had a flashlight. As he got within twenty feet of it, he recognized it as a human and sprinted the rest of the distance to its side. It was a man, naked except for a twisted hospital gown wrapped around his neck and head and one yellow sock. Mix searched for a pulse and found the man to be alive, albeit barely. He sprinted back to his car, dialing 911 as he ran.

"911, what's your emergency?" came the instant response.

"I...I...I found a man on the side of the road, and he's barely alive," he said as he opened the back hatch of his car and grabbed a blanket he always kept there for his children.

"Sir, could you please slow down and repeat yourself?"

"There's a dying man on the side of the road! He's barely alive! I'm about two miles north of the City of Ithaca on route 96!"

"Sir, there's a lot of background noise, and I'm having difficulty understanding you. Did you say there's a bear?"

"OH MY GOD LADY! I SAID... oh fuck it!" he said, tapping off the phone and sliding it into his pocket. *The hospital is right up the road here, a couple of miles,* he thought as he wrapped the man in the blanket, lifted him, and carried him to the car, placing him supine in the back of the Subaru wagon. He was taken by how light and frail the man seemed. On the way to the hospital, he called 911 again and announced his soon arrival at the ER so they would be prepared.

When he arrived, he parked directly in front of the emergency room entrance and retrieved Adams from the back of his car. He was promptly met by two male nurses with a gurney who took Adams from Mix and gently placed Adams on it. Mix followed them in and gave a full report to the attending physician on duty.

# CHAPTER THIRTY

**Washington, D. C.**

They sat at the end of the large, heavy, dark mahogany conference table in maroon leather chairs in the meeting room on the third floor of the federal building at 935 Pennsylvania Avenue NW, Washington, DC 20535. Sandy was at the head of the table. To her right sat FBI Special Agent Tom Foster. Across the table from Foster was FBI Special Agent Mary Broadworth. Mary was a ten-year veteran of the FBI, following in her mother's footsteps, who'd retired as an FBI sharp-shooter agent just three years ago. Mary was always dressed as if she was about to walk into the courtroom to take the stand to testify—navy blue suit, skirt cut just above her knees, one-inch black pumps, and a crisp white pressed shirt. Other than that, her appearance belied her power; she was just five-foot-one, slim and fit.

"Tom, I know Upstate New York is your assigned territory, but the U. S. Attorney in Binghamton has agreed to assign you to me on special assignment," Sandy said.

"Right. That's why I'm here, Sandy."

Sandy turned to Mary.

"I see you've got the file with you, Mary."

"Yeah. It's what you asked for. The forensics lab was fast."

"And?" Sandy asked.

"His real name is John Holdings. DNA clear as a bell."

"Priors?" Foster asked.

"Arrests, no convictions. Here, let me tell you," Mary said, opening the manilla file in front of her. "Three DWI's. First one, a conditional discharge, the second—that would've been a felony—dismissed. And the third one, just dead-ended."

"Dead ended?" Sandy asked, puzzled.

"Dead-ended," Mary said. "No discovery, no court appearances, no prosecution at all. Just a dead-end."

"Interesting," Foster said, scratching his head.

"So, our 'Volkov' is no Russian at all," Foster said. "Where's he from?"

"Interestingly enough," Mary said, "Annapolis. We traced him through the National Coast Guard and records."

"Because of the boat accident on the Potomac?" Foster asked.

"No, because he's actually a licensed captain—'Six-Pack.'"

"What's beer got to do with it?" Sandy asked.

"No no no," Mary said, a big smile across her face. "It's a U. S. Coast Guard Six-Pack captain's license that lets the captain take no more than six paying passengers and crew out on his or her boat."

"Ah," Sandy said. "NTSB OMS?"

"National Transportation Safety Board, Office of Marine Safety?"

"Yeah," Sandy said.

"Nope."

"No?"

"No, Sandy, the OMS only investigates major marine casualties—six or more deaths or the loss of a vessel one-hundred tons or more. He was questioned by the NTSB 'Go Team.'"

"Results?" Sandy asked.

"Got 'em right here," Mary said, flipping though her file. "Each questioning ended with the exact same conclusion: 'interviewee had little to no recollection of the cause of the accident.'"

"That's fishy," Foster said. "No pun intended."

"Very funny, Tom," Sandy said, shaking her head.

"Oh, there's more," Mary said. "He's been brought in for questioning in three marine fatalities, not counting the one he posed as 'Volkov' in."

Mary flipped through the pages of the file in front of her.

"Uh, yeah. Here," Mary said. "Marion Miller. Accidental drowning at the Bay Sound Marina, Annapolis. Apparently slipped off the dock as she was boarding a newly-acquired vintage 1929 seventy-five-foot Elco yacht with her husband, struck her head on the dock as she fell. Husband John Miller." She flipped the page. "Her estate settled by attorney Randolph Everts. She died with a three-million-dollar life insurance policy on her. Husband sole beneficiary."

"And?" Sandy said.

Mary flipped paged, "Holdings was apparently the 'hero' in this one."

"Hero?"

"Apparently, Susan Spaulding was rowing her dinghy in the marina at night, lost her balance and fell in. Holdings jumped in to save her but lost sight of her in the dark."

"I see."

"He did retrieve the dinghy."

"Of course," Sandy said.

"Toxicology report indicates her blood alcohol content, BAC .36."

"That's near-fatal," Sandy said.

"Yup," Mary confirmed.

"That place…I don't know," Sandy said. "The others?"

"Sure. Virginia Bennington, 'lost at sea' during a party on a vessel named '*AVENGER*.' Name of the boat later changed to '*MY CHOICE*' by her widower, Victor Bennington."

"*MY CHOICE*, huh?" Foster said.

"Yes," Mary said. "And now I'm glad you're all sitting down."

"Why?" Sandy asked.

"Well, I traced the Coast Guard Documentation records on that boat. These larger yachts aren't usually registered under the state because they travel on the oceans, so they're documented by the Coast Guard."

"And?" Sandy asked impatiently.

"Last known owners Vince DeMarco and our very own Mary Beth Allison."

Tom and Sandy sat in stunned silence. Mary closed the manilla folder and slid it over to Sandy.

Finally, Tom spoke up. "Well, that certainly brings us full-circle now, doesn't it?"

"This links the Weber murder to the suspicious deaths at the Bay Sound Marina, does it not?" Sandy asked.

Foster nodded.

"It does, Sandy. It does," Mary confirmed. "And Holdings, our 'Mr. Volkov' is the link."

The room fell silent as they all took a moment to let it all sink in.

Sandy picked up the folder Mary had slid over to her on the table and opened it, turning through the pages until she found the last U. S. Coast Guard Document. Sure enough, Vincent DeMarco and Mary Beth Allison were documented owners.

She picked up the phone on the conference table, looked up Mary Beth's number on her cellphone and dialed. It went to voicemail. She hung up the receiver. "Voicemail," she said, then looking at her watch.

"I've gotta go. I've got a suppression hearing with Judge Mayfield downstairs. I'll try Mary Beth later," she said, holding up the blue and white Coast Guard Documentation for *MY CHOICE.* "Tom, you follow up with her."

"Will do," Foster said as they all stood to leave the meeting.

As Sandy strode down the hallway toward the courtroom, she dialed Mary Beth again.

***

Vince and Mary Beth were enjoying their usual breakfast of avocado toast and a cup of fruit salad and coffee. They sat close to each other at the small round table overlooking the view of the lake, although, with the rain and haze, there wasn't much of the lake they could see.

Mary Beth's phone sounded: "Unknown caller"

She and Vince both looked at the phone, ignored it, and went back to their breakfast. A moment later, the phone pinged, indicating someone had left a voicemail. They continued to ignore it. Finally, the curiosity got to them, and Mary Beth picked up the phone and listened to the voicemail. She tapped off the message and slowly put the phone down on the table, her eyes meeting Vince's.

"My God, Mary Beth, what is it? You look like you've just seen a ghost."

The message was short and succinct; it was Assistant United States Attorney Sandy Mitchell in a low, serious voice telling Mary Beth to contact her immediately and "don't waste any time." She put the phone on "speaker" and played it for Vince.

"That's the U. S. Attorney I worked with in D. C.," she explained.

"Well, for God's sake, call her!" Vince said, "Call her!"

Mary Beth tapped the call-back tab, and the phone dialed Sandy Mitchell's personal cellphone. She answered on the second ring.

"Hello, Mary Beth," she said in an almost somber voice.

"Hi, Sandy. What's going on?"

"Well, there's not much I can disclose to you right now, but I have to tell you we have intelligence that you may be in some danger."

Vince could hear Sandy's voice on the phone, and his chest tightened.

"Danger?"

"It's nothing new in our business, Mary Beth. You know that."

"I do, but can you give me details?"

"I can only tell you that it is related to one of your old cases down here and that we are placing resources for your protection."

"Do I need to come in?" Mary Beth asked.

"No, not yet. If it comes to that, you know I'll send for you."

"OK."

"And Mary Beth, don't worry, but stay alert like you always were when we worked together."

"Got it."

"I'll stay in touch," Sandy said.

"I appreciate it."

"You bet. Take care."

The call ended. Mary Beth turned to Vince.

"My God, now *you* look like you've seen a ghost!" she said.

"What's going on, Mary Beth, and why are you in danger?"

Mary Beth had composed herself, feeling the FBI in her coming alive, and liking the feeling.

"No worries, Vince. Standard Operating Procedure for ex-FBI agents. We make a lot of people very very angry when we arrest and prosecute them. The ones who threaten are the cowards and nothing to worry about. It's the ones who say nothing that are the real danger."

"Someone threatened you?"

Mary Beth took a moment to think. Sandy never said someone threatened her; she only said they had intelligence she was in danger. That's not the same. She knew what Sandy meant with those very precise words. She was in serious danger.

"Sure. Probably just some coward I arrested years ago. Nothing to worry about, Vince."

He knew better.

***

Sandy strode out of the federal courtroom on the first floor of the federal building, her file tucked under her arm. She had just finished arguing against a motion with a defense lawyer who had filed a motion to suppress three kilos of cocaine as evidence. He told the judge the search by the FBI had been illegal because the cocaine was found in a closet, in a duffle bag, in a garbage bag, in his client's bedroom and was therefore outside of the purview of the search warrant the judge had signed allowing for a search of the basement. Judge Mayfield did not seem convinced but reserved his decision.

As she walked down the white marble-floored hallway, she dialed Tom Foster on her cellphone.

"Tom, have you reached out to Mary Beth yet?" Sandy asked.

"Just got off the phone with her," he confirmed.

"And?"

"She said she'd spoken to you briefly and was expecting to hear from me. We don't have to subpoena her; she's more than anxious to help."

"I'm not surprised one bit," Sandy said.

"Oh, and I spoke to detective Ken Ludlow, up in Ithaca too. He's onboard."

"Good," Sandy said, with no pause in her stride. "Set up the meeting."

"Got it."

"And go get Holdings."

"Wilco."

"And Vince DeMarco."

***

At midnight, the winds were howling from the Northeast at thirty knots, and the swells had reached four-to-six feet on the Chesapeake Bay, six-to-nine feet as the swells rolled into the shallows, as the one-hundred-foot, former presidential yacht, *EMANCIPATION,* pulled into the marina, her one-hundred-ton hull heaving violently in the crashing waves. The captain and three crew members struggled to try to dock her and tie her up.

Rick Anders stumbled up the pier toward the boat to help. He'd been drinking heavily that night—more than usual with the stress of knowing Bryce and his boat were coming—and made his way to the end of the dock. As he approached the yacht, a crew member tossed Rick a line from the bow of the boat, and Rick reached out to grab it, barely within his reach. He managed to grasp it with his left hand, but when he tried to get control of the line with his right, he lost his balance and was pulled into the water between the dock and the yacht. In the dark and with the heaving hull, howling winds, growling engines and pounding, slamming waves, no one saw or heard him fall.

Twenty minutes later, the captain and crew finally managed to dock the boat. She was at the only place in the marina that would fit her: the "T" dock at the very end of the pier, adjacent to and within twelve feet of *MY CHOICE.*

Rick was nowhere to be found as the strong undercurrents swept him out into the deep. The crew assumed he'd given up and left.

# CHAPTER THIRTY-ONE

**Ithaca**

Detective Ken Ludlow had just returned to his office after concluding his part of an investigation into a two-car fatal accident, alleged drug involvement, and the discharge of a handgun that set off the entire chain of events when his intercom buzzed.

"Yeah," he said.

"Got the hospital on line three," the office receptionist announced.

"Alright, I got it," he said as he picked up the receiver and punched the line three button. "Detective Ludlow, " he said flatly.

"Detective, this is Doctor Menton at the county hospital."

"Yeah, doctor, what can I do for you?"

"Well, I've got a patient just admitted that I've been told you were up here trying to see, and he's back. Sean Adams."

"Back? What do you mean, 'back'? Wasn't he on a seventy-two-hour hold?"

"Well, no matter. He's here if you want to speak with him, but I have to tell you he's in bad shape—critical but stable."

"What happened to him?" Ken asked as he stood and grabbed his patrol car keys off his desk.

"Don't know yet; can't say, Detective. HIPPA privacy rules. You know that. All I can say is he was found on the side of the road."

"The side of the road? HIPPA, my ass. I'm on my way," Ken said as he slammed the receiver down and hiked to his car.

***

Big Red had left Adams for dead after feeling his neck snap as he choked him up against his SUV. What he had done was cause a traumatic hyoid bone fracture in his neck—painful and dangerous but not fatal. The greater danger had been hypothermia and the risk of being struck by a passing car as he lay on the side of the road unconscious.

Ken called District Attorney Becker on his way to the hospital.

"Gary, you won't believe this," Ludlow said.

"Oh, try me," Becker said.

"Apparently, Sean Adams somehow escaped the seventy-two-hour hold, and he's back in the hospital but in critical condition."

"No shit."

"Yeah. I'm on my way to the hospital to see him now. You want to meet me up there?"

"Sure. He's a member of the local Bar. We owe him that much. Hope he's OK."

"Well, that's all well and good, but we still need to talk to him about his involvement with Alina Weber."

Becker took a few seconds to think. "I think that ship has sailed, my friend. I turned everything over to the Feds."

"Well, I still want to talk to him about it," Ken said, knowing he was to meet with the U. S. Attorney in D. C. Soon.

"That's up to you, Ken. See you up there." He clicked off the call.

***

Adams was still in the emergency department and hadn't been sent upstairs to a hospital room yet when Ludlow arrived. He was resting comfortably with a morphine drip for the pain.

"Jesus, Adams, you look like shit!" Ken said, stunned by how gaunt Adams appeared.

Adams tried to respond, but the pain was too much, so he just raised his hand in a weak wave to Ken.

"Looks like you had quite a night."

Adams forced a meager smile, gave a 'thumbs-up', then summoned Ludlow toward him with a hooking index finger gesture.

Ken approached Adams, and Adams repeated the gesture until Ken bent over and was inches from him.

Adams tried to speak.

"I..."

He coughed and grabbed his neck in pain, then raised his index finger to Ken, signaling, 'give me a second.' Ken nodded.

"I...I..."

"Yes?" Ken said, getting impatient.

"I...I...k...k...killed Weber."

With that, Ludlow stood back and started laughing.

"You *what?* You're hallucinating, Adams."

Adams slowly shook his head no and repeated, "I...k...killed...Web...Weber."

Just then, Doctor James Menton walked into the bay and, seeing the look on Adam's face, turned to Ken.

"OK, that'll be enough," he said, "he's got to rest. So, if you please."

Ludlow was baffled and, vowing to himself he would be back, turned and began to leave the bay just as Gary Becker appeared in front of him. He took Gary by the elbow, turned him around, and they walked out of the ER and to the hospital lobby before another word was spoken.

"You're not going to believe this, Gary."

"What again? Try me."

"Adams just confessed to killing Weber."

Gary burst out laughing. "Oh, that's the funniest thing I've ever heard!" he exclaimed, "That's fucking hilarious! That guy couldn't hurt a fly."

"Well, that's kinda what I thought, too. And he is on morphine right now."

"That explains it, doesn't it? And for Chris sakes, we saw him have a nervous breakdown right in front of us, did we not?"

"I guess. But I want to know where he was and how he got into this condition."

"Sure. We've still got jurisdiction over what may have happened to him if there's any possible criminal activity behind it, aside from the Weber murder. You do your digging; let me know what you find. And don't do anything to interfere with the federal Weber investigation."

"I wouldn't think of it," Ken said with a smirk, knowing he was to be a part of that investigation.

"Good," Gary said. "It's out of our hands now."

Ken paused.

"Gary."

"Yeah."

"You think there's a chance Adams killed Weber?"

"Ha. Not a fucking chance. He's full of shit."

"Right."

"Yeah. I wouldn't waste another minute on it. Let's get out of here. This place depresses me."

***

Vince was in his car headed for the office to confer with Michael on filing a notice of appearance with the U. S. Attorney's Office and Federal Court in the new federal Weber case when his phone sounded through the car's Bluetooth system, and an unknown number appeared on the Tesla's screen. He ignored it. Then the same number called again. He continued to ignore it. But the third time that number popped up on the screen, he tapped the "accept" tab on the screen and answered the call.

There was a faint voice, a whispering on the line that Vince couldn't make out. It gave him a chill down his spine, and he figured it was a scam or prank call, so he tapped off the call. A minute later, the system notified him there was a voicemail. He figured it couldn't hurt to listen to it, so he tapped it on.

"Vince, this is Sean Adams. I need you. I need you bad. Please call me back," the whispering voice said.

*Damn!* Vince thought. *If that was really him, he sounds like shit.* He tapped 'call back' on the screen, and it dialed Adams back. He answered on the third ring.

"Vince, thank you for calling," Adams said in a barely audible whisper-voice.

Vince cringed. "Sure, Sean, what's up, and what's wrong with you?"

"I need you to do me a favor, Vince, and it's critical."

"I'm listening," Vince said, dubious he would agree to anything.

"My car."

"Your car."

"Yes. I need you to get into my car."

'You want me to get into your car," Vince said flatly.

"Yes. I don't know where it is, but it may have been towed from a handicap spot behind the courthouse."

"You parked in a handicap spot behind the courthouse? Then it's definitely been towed. What do you expect me to do?" Vince asked, thinking, *what an idiot.*

There was a long silence on the line as Adams tried to swallow through the pain in his throat.

"Adams?"

"I don't know. All I can tell you is that you need to get into my car, it's a red BMW. When you do, you'll know exactly why. And it's for Alina. Look, I'm not good right now, morphine, you know. Gotta go."

The call ended.

Vince called Detective Ken Ludlow without missing a beat.

"Vince. What's up?"

"Ken, where would a car towed from behind the courthouse be taken?"

"They got your Tesla, Vince?" Ken said with a chuckle.

"Nope. I'm driving it right now. So?"

"Sure. They'd call Fulton Street garage. They do all the towing down there. If they got it, it'll be in the lot out back of the repair shop."

"Thanks, Ken. Much appreciated."

Vince ended the call, then redialed Adams.

"I think I know where your car may be, Sean. But I don't know where your keys are."

Adams took time to think.

It came to him suddenly. He'd never locked his office when he bolted out of there to get away from Big Red, and his keys were probably still on his desk.

"Vince, my office is unlocked, and my car keys should be right on the desk."

"Alright, and I have consent to go into your office?"

"Of course."

"And this is for Alina's sake?"

"Absolutely."

"You want to give me a clue?"

The call went dead.

"Damn!" Vince said aloud. "Damn!"

Instead of continuing to his office, Vince went straight to the Circles Building and to Adams' office. Just as he said, the keys were on his desk, among a mess of papers. As he reached for the papers, he noticed the death certificate for Dr. Weber, then the life insurance policy, and the Will. He paused and, after a minute, scooped up the life insurance policy, put it in his inside jacket pocket and left the office, heading to the Fulton Street Garage. He called Michael en route.

"Michael, meet me at the Fulton Street Garage. You know where that is?"

"Sure, Dad. What's up?"

"I have no idea. I just spoke to Sean Adams. I have his car keys, and he says I need to get into his car for something."

"For what?"

"No idea. Just meet me there."

Mike shook his head. "Sure, Dad. Whatever you say."

Twenty minutes later, a Code Blue alarm went off at the hospital. Sean Adams was in cardiac arrest. He would not survive.

***

Vince and Michael met at the Fulton Street Garage, went directly to the office counter, and inquired as to a red BMW.

"Sure. Red BMW. We got it. It yours?" The man behind the counter asked.

"No, we're attorneys, and we just need to inspect the interior of the car."

"Lawyers, huh? Never been my favorite people."

"Until you need one," Mike said, turning to Vince with a wink.

"Yeah, sure, kid. Alright. You got the keys and some kind of ID?"

Mike pulled out his court ID card and set it on the counter.

"Oh! DeMarco. And you must be the old man," he said, looking at Vince.

"Yup. I'm the old man for sure," Vince said.

"Well, OK. One of you guys represented my sister for her divorce. You done good. She got the kids, the house, alimony and child support and half the jerk's retirement. You done good. OK, this way," he said, pointing to a side exit from the building. "You can't miss it. It's right there," he said, pointing. "Have fun."

The car was covered in damp leaves and dirt but otherwise intact. Vince pressed 'unlock' on the FOB, and the car unlocked. He opened the driver's side door.

"Well, Dad. You've probably just tampered with evidence," Mike said.

Vince ignored him as he leaned into the car, scanning for anything that might mean anything. He saw nothing.

Mike peered through the tinted driver's side rear window.

"Dad."

"Yeah, Mike."

"Check the back seat."

"Alright," Vince said, straining to peer over and between the two front seats.

"See something?"

"Yeah. I see a rubber glove."

"Don't touch it!"

Vince leaned back out of the car, turned to Mike and said, "Duh."

"Alright, sorry, Dad."

Ten minutes later, Detective Ken Ludlow was on the scene with a forensic officer in response to Michael's phone call.

"You say this is Sean Adam's car?" Ken asked Vince.

"Don't know for sure, but the registration is probably in the glove box. He told me it is."

Ken and the forensics officer scoured the entire interior of the car, took prints inside and out, and located the two rubber gloves Adams had tossed in the backseat after the shooting. Ken went into the garage office and arranged to pay the towing and storage charges and have the car towed to the sheriff department's secure lot behind the offices.

"You get your hands all over everything in there, Vince? We going to find your prints all over the place?"

"Very funny, Ken, very funny. Just the door handle."

Ken's trained eye immediately noticed minute dried blood spatter on one of the gloves. He knew they'd probably find GSR—gun shot residue—as well. Now, he believed Sean Adams.

*Unless we find Alina's DNA inside those gloves,* he thought as he walked back to his car. *What the fuck.*

# CHAPTER THIRTY-TWO

## Annapolis

John Miller, owner of the 1920 Elco, Captain Randy Spaulding, owner of the 1930 Defoe Commuter yacht and Bill Brown sat in a semi-circle around the presidential seal woven into the ancient carpet on vintage blond mahogany and dark green leather chairs across from Ron Bryce in the main salon of *EMANCIPATION.* Bryce had a Monte Christo cigar in hand, just lit. The salon was made of the finest teak and dark mahogany, trimmed in 14-karat gold, surrounded by low cabinets with stained glass and gold-trimmed doors. The heavy hunter-green drapes were pulled closed.

"So, where is he?" Bryce asked Bill Brown.

"Oh, he's around somewhere, Ron. Probably be here soon."

"You know, the main reason I spend so much time and so much fucking money to cruise my boat up here to this miserable, dilapidated pathetic marina is for him to do my annual service. So why isn't he here? You knew damn well when we'd be pulling in."

Bill, John and Randy were truly puzzled. Rick had never gone missing—not even for a day before, and they knew Bryce was no one to mess with.

"You either produce him, or I swear I'll burn this fucking place to the ground, understood?"

They all nodded, not doubting him for a second.

"Now, other business," he demanded.

Bill cleared his throat. "We have a slight problem with the Matthews," Bill said.

"Problem?" Ron said, leaning toward Bill.

"Slight."

"And what is the *slight* problem?"

Bill glanced over at Captain Randy Spaulding.

"Well, Randy here was supposed to cruise it out of here when we sold it, but the new owners decided to keep her here. That's the Matthews docked right next to you here."

"Wait!" Bryce said, "You mean the Bennington boat is still here? Right here? Are you insane?"

Suddenly, Bill felt a shot of panic as he remembered that Rick had made that pinhole in the Matthews tank, dripping diesel into the bilge, which was now within twenty feet of *EMANCIPATION.* He'd have Rick fix that first thing when he showed up from wherever he was. *That's too close for comfort,* he thought to himself.

Spaulding turned to Bill. "Tell him, Bill."

Bryce glared at Bill, and Bill took a deep breath.

"There's more," Bill said, dreading the next minute of his life.

"More," Bryce said flatly, his glare intensifying.

"The new owners. One of them is a former FBI agent."

Bryce threw himself back in his chair and started laughing. The others chuckled too until Bryce's laughter became maniacal; they all stopped and waited for him to stop; it went on for an abnormal, sick, long time. He finally stopped, composed himself, and wiped the tears from his eyes.

"Oh, you guys!" he said, chuckling. "You're killin' me!

They all smiled.

In an instant, Bryce stood, still smiling, drew his Baby Browning A .25 ACP Pocket Pistol from his right front pants pocket and put a bullet in John Miller's forehead. The small caliber pocket pistol didn't kill him instantly. He put his hand to his forehead with a confused look on his face, then slowly slumped backward in his chair—his eyes wide open.

Brown and Spaulding sat paralyzed, staring at Miller.

"Alright, gentlemen," Bryce said. "I hope you get my point. So, just so you know there's no hard feelings, what I want you to do is, get me Rick Anders so he can start to work on my boat tomorrow, get that fucking death boat Matthews out of this marina and to the bottom of the Chesapeake, and that FBI agent as dead as your friend here is. Then guess what? Then maybe I'll give you another chance. Understood?"

They both nodded.

They both nodded nervously.

"Now go get a fucking dock cart, and get him the fuck out of here! NOW!"

The two scrambled off the boat and semi-sprinted down the pier to retrieve a dock cart.

"Oh, Jesus!" Spaulding said. "Jesus! What do we do?"

"Let's kill two birds with one stone," Bill said.

"What?"

"I said, let's kill two birds with one stone."

"Oh. Holdings."

"Yeah. Holdings."

"Got it," Spaulding said.

"But that means…"

"Yeah. We have to sacrifice Holdings' 1930 Frank Anderson Yacht."

"What a pity," Bill said, "such a pity."

"So Holdings and Miller go to the bottom in the Frank Anderson," Spaulding said, shaking his head.

"It's how they'd want it to be," Bill said somberly.

"So true. So true."

***

FBI Special Agent Tom Foster, FBI Special Agent Forensics Alfred Platford, the FBI blood-spatter analyst and Sandy Mitchell, Assistant United States Attorney, met in Sandy's office in the D.C. Federal building. They each had copies of the entire investigation conducted under Gary Foster, Tompkins County District Attorney's authority up in Ithaca, New York. The package of discovery materials Foster was just about to hand over to Michael DeMarco before the Feds intervened in the case.

"No GSR, not a speck of gunshot residue anywhere on Alina Weber. Nothing—not her hands, fingers, wrists, chest, neck, clothes—nothing," Foster said.

"The way we see it, Sandy, there's no way she pulled that trigger," Platford added with confidence. "And as you can see from Deputy Wilson's report, he was on the scene so fast, the gun was still hot. She didn't have the chance to clean up."

"What I need to know is where did the blood came from, how was the perp positioned, what movements did the shooter make after the shooting and could there have been more than one other person in the house besides the shooter and the victim," Foster said.

"Let's go over the photos again," Platford said.

They all pulled the packet of eight-by-eight-by-eleven-inch color photos out. Platford spread them out across the top of the table. He was well rehearsed for this, having already gone over it with DA Foster.

"So what we look at is forward spatter and back spatter. Forward spatter is blood that spatters from the exit wound, and back spatter that comes from the entrance wound."

"We know that," Foster said, knowing exactly where this was going.

"Well, just follow me on this," Platford said.

"Don't interrupt, Tom," Sandy admonished.

"Alright," Platford continued. "All that depends on the caliber of the gun, where the victim was located, whether the slug lodges in the body or not, how far the perp was from the victim and where the victim was relative to the environment—floor, walls, and other objects. Forward spatter is like a mist, and back spatter is larger droplets."

"OK," Sandy said, "and?"

"Pull out the photo marked 'Exhibit 4." Platford said.

Sandy grimaced. "Ugh, that's horrible," she said.

"Yeah," Platford said, "Weber's body slumped over on the couch. This photo was taken from exactly two feet from the victim. Do you see the spot marked by a red marker?"

"Looks like a bullet hole," Sandy said.

"Exactly. Only one bullet passed through the victim."

"But there were three gunshots, yes?" Sandy said.

"Yup. Two were lodged in the occipital bone, In other words, the very bottom of the back of the skull."

"Alright, tantalizing enough, get to the point," Foster said.

Sandy and Platford exchanged glances.

"Now the back spatter," he continued. "Pull out the photo marked 'Exhibit 6'."

Sandy pulled her copy out, and carefully examined it. "Someone was standing between the victim and the wall directly across from him when the shots were fired. There's clearly a spot, almost in the shape of a person, where the Luminol didn't react on that wall with the back spatter."

"Exactly, Sandy," Platford confirmed.

"Well, that person would've been covered with back spatter blood, at least as much as the wall behind him or her was, and GSR."

"OK, so pull out 'Exhibit 7'," Platford said.

"Oh, this is my favorite," Foster said. "I just love computer-generated graphics."

Sandy crinkled her nose, not sure whether Foster was being sarcastic or not.

Platford pulled out his copy and paused. "You got it, Tom?"

"Yeah. I got it. Go on."

"Thank you. Based on the one bullet hole in the back of the couch and the two bullets lodged in the lower edge of the occipital bone, we know the angle of the three shots. First, we can tell they were all fired from the same distance and angle."

"And in rapid succession," Tom Foster added.

"Yes," Platford said. "We know they were fired in rapid sequence because the body of the victim didn't move much from one shot to the next. They're close in a cluster. There was no pause between shots at all."

"OK, and?"

"'Exhibit 8,'" Platford said. "Take a look."

Foster was getting impatient but knew Platford had to take Sandy through the exhibits one at a time.

"It's a photo of the floor between the front door and the couch," Foster interjected. "Check out the pattern in the Luminol, Sandy."

"Yes. And note the distance to the victim marked in scale," Platford said. "sixty-four inches from the Victim's feet on the floor at the couch to the two spots not reactive to the Luminol. So what's that tell you?"

"From the angle of the shots, the area on the wall not showing blood spatter, and the two footprints on the carpet that showed up after spraying the Luminol, the shooter was almost on top of the victim, or at least the gun was," Sandy said. "And the shooter had to be tall, probably six feet two at least, to have that reach."

She took some time to look over the photos again and the computer-generated exhibit. She placed them back in the folder in front of her on the desk, leaned back in her chair, crossed her arms in front of herself, and said, "Was Gary Becker really going to take this to trial?"

"Last we knew," Foster said.

"Well, that's just ridiculous. There's no way in hell that girl shot her husband. No gunshot residue or back spatter on her, and now this." She shook her head.

"So, now what?" Foster asked Sandy.

"I'm going to have a little talk with District Attorney Gary Becker," she said. "Thank you, gentlemen. Good work. I'll keep you in the loop."

The meeting ended, and Foster and Platford got up and left Sandy's office.

"Close the door behind you," she said.

"You bet," Foster said as he left.

She dialed Gary Becker's number, and the receptionist passed her through to him.

"Gary, Sandy Mitchell here. How you doing?"

"Oh, alright, I guess, Sandy. What can I do for you?"

"I think it's more what I can do for *you,* Gary."

"Oh really?"

"Yes, really. We've been over everything you sent us on the Weber murder, and I'm going to make a motion that the indictment be dismissed and charges dropped against Alina Weber."

"Very interesting, Sandy, but I'm not surprised."

Gary's intercom sounded, "Gary, Detective Ken Ludlow just stormed in and headed to your office; I couldn't stop him."

Gary looked up, and Ken was standing in front of his desk.

"Whoever you're on the phone with, Gary, hang up," Ken demanded.

"Hold on, Sandy. I'm going to put you on hold for a second. I'll be right back," Gary said, punching the 'hold' button on the phone and putting the receiver down.

"What the fuck, Ken!"

"What the fuck is right, Gary. Glad you're sitting down."

"Well, have a seat yourself, and tell me what this is all about."

"DeMarco just completely slammed the murder case against Alina Weber into the ground."

"How?"

"He tracked down Sean Adam's BMW and called me to the scene. We found two rubber gloves thrown in the back of the car."

"And?"

"I sent them to the lab, and there was blood spatter matching Dr. Weber's DNA on one of them and Sean Adams's DNA inside the gloves."

"They had Adam's DNA samples?"

"They matched it to DNA on a paper coffee cup in the car."

"No shit," Gary said, feeling a little dizzy.

"Oh, there's more, Gary. Yesterday, I went up to the hospital to see him, as we said, and he confessed to killing Weber."

"And we did nothing?"

"Nothing. I figured he was still out of his mind and probably hallucinating."

"But the glove."

"I went up and saw him before I even knew about the glove."

They sat pondering for a couple minutes.

"That may clear Alina Weber of the shooting, but it doesn't

clear her if she was part of a conspiracy to kill Weber," Gary said, clearing his mind.

"You're right. There's still that. What proof do we have of that?"

Gary stroked his chin. "Hm. Nothing."

Gary's intercom sounded again.

"Gary, Attorney Sandy Mitchell is on line two. You hung up on her, and she's calling back."

"Damn," Gary said, "I never have learned how to use these goddamn antique telephones." He picked up the receiver.

"That was rude, Gary," Sandy said.

"Oh, sorry, Sandy. I punched the wrong button on the phone."

"Sure you did, Gary. So listen, I have a proposal for you on the Weber case."

"Wait, Sandy. I've got something to tell you. I just found out this minute. We have the murderer in that case, and it is not Alina Weber. At least she wasn't the actual shooter."

"Well, then, I guess that makes my call moot. I was about to tell you to either dismiss the indictment or I would be filing a motion to get rid of it."

"I wouldn't move too fast on that, Sandy. There is the distinct possibility Alina Weber and Sean Adams conspired to kill Weber together."

"Sean Adams? Who the hell is he?"

"We think he may have been her lover, and we have traced the murder weapon back to him and one other possible suspect."

"And you're telling me this just now," Sandy said, clearly irritated.

"I promise you, Sandy, I just found out about Sean Adams as the shooter between these phone calls with you."

"I'm going to ask Judge Britton for a conference. We've got a lot to talk about, Gary."

"That's fine," Gary said. "I'll set it up," he said, ending the call. He dialed the DeMarco Law Firm and got Michael on the phone.

"I've got some new for you, Mike," Gary said.

"Good or bad?" Michael asked.

"Pretty good. We're dropping the murder case against your client, Alina Weber, for now."

"Well, that *is* good news...but wait. What do you mean, 'for now'?"

"The gloves we took from Sean Adams' car had blood spatter on the right one; DNA matched Dr. Weber. DNA on the inside of both gloves was from Sean Adams. That makes him the shooter."

"Sure it does."

"I'm just not sure there wasn't an agreement between Alina Weber and Sean Adams to kill Weber."

Mike gave a chuckle. "A conspiracy to commit murder? You must be kidding!"

"I'm not, but I don't have proof yet, so we're going to go ahead and drop the indictment against her, but keep her close, Mike. As soon as I can make the connection, we'll be arresting her again. Meanwhile, she's all yours."

"I'll take this as all good news, Gary, because there's not a shot in hell she was involved in a conspiracy with Adams."

"What other motive would Adams have?" Gary asked.

Mike paused.

"I don't know, Gary; I just don't know."

"OK," Gary said, "you go ahead and file a 210.20 motion to dismiss the indictment, and I'll advise the judge we have no opposition to it. You get to be the big hero."

"Fair enough, Gary. I'll file it today."

Mike hung up and promptly called Vince.

"Congratulations, Dad, they're dropping the indictment

against Alina based on the gloves they found in Adam's car. Blood spatter on the outside of the right glove matched Dr. Weber, and DNA in the gloves matched Sean Adams. Makes him the shooter. You did a good job."

"Thanks, Michael."

"Only one thing, though."

"Yeah?"

"Gary's got the notion Adams and Alina might have been in a conspiracy to kill Weber."

"Hm. Actually, not out of the realm of possibility, Mike. What other motive could Adams have had to do the murder?"

"Oh, I don't know, maybe pure blind hatred of the guy."

"Or maybe he and Alina were lovers," Vince suggested. "But whatever Gary's theory might be, I know him enough to know he won't make another move without real hard evidence."

"I agree. We need to give Alina the good news," Mike said.

"I'll give you that honor, Mike. After all, I'm only second chair on the case!"

"I appreciate it, Dad. And we need to talk to Adams if he's not represented by counsel yet."

"Sure, but if he asks, we cannot represent him. Conflict of interest," Vince said.

"Duh."

# CHAPTER THIRTY-FOUR

**United States Federal Building**
**Binghamton, New York**

The U.S. General Services Administration (GSA) describes the United States Federal building in Binghamton, New York, as a building with an "austere and temple-like facade," with seven three-story columns supporting a "continuous Greek key frieze," built in 1935 in marble and limestone. It is, by all definitions, an awe-inspiring, if not fear-inducing, building. The meeting would be here to avoid making the "invitees" travel to D. C.

Sandy Mitchell, Assistant United States Attorney, had reserved the large conference room on the first floor in the U. S. Marshal's quarters. She sat at the head of the ten-foot, marble-topped, dark-mahogany conference table, a closed laptop in front of her; Mary Beth sat to her right, Vince DeMarco next to Mary Beth, and across the table sat Michael DeMarco to his left, Alina Weber, to her left, Special Agent Tom Foster, and at the end of the table, Detective Ken Ludlow.

As usual for a meeting of this type, the U. S. Marshals had searched and scanned the room for hidden weapons, micro-explosives and listening devices., And had set up a fresh pot of coffee, a bowl of creamers, packets of sugar and a dish of doughnuts on a sideboard against the wall. No one touched any of it.

"Well, first, let me thank you all for coming," Sandy said gratuitously, looking around the table.

"Did we have a choice?" Mike said.

Sandy smiled at him. "There's always a choice, counselor," she said.

Vince gave Mike a stern look.

"We have a lot to cover, so let's get to it, shall we?" Sandy said.

"Sure," Mike said. "Alina's here based on your subpoena, so may we start with her?"

"Of course," Sandy said.

Mike slid a sealed eight by eleven white envelope across the table to Sandy. She picked it up and opened it.

"Ah, good," she said, sliding out the contents: a single sheet of yellow-lined paper with four phone numbers handwritten on it. One of them was Alina's mother's number; another the Russianbryde's emergency number; the third Jason's and the fourth, Michael's.

"That's it?" she asked Mike.

"You have her phone," Mike retorted.

"That's true. We also have Dr. Weber's phone, his laptop and desktop computers."

"It's Ok, Mike. We just needed to know if Alina had any contacts that were not in her phone."

"She hardly used it," Mike said. "She wasn't allowed."

"So I understand," Sandy said. "Do you mind if I ask Alina some questions, Mike?"

"Go ahead."

Sandy opened her laptop and scrolled with her index finger.

"Russianbryde," she said, looking up at Alina.

Alina immediately blushed bright red. "Yes?" she asked.

"What can you tell me?"

Alina looked at Mike, waiting for him to say something to stop the question.

"Answer her," he said.

"It's how my husband found me."

"It's a mail-order bride service, is it not?" Sandy said.

"Mail?" Alina asked, looking at Sandy and then at Mike.

"Sandy, we know what it is. What's your point?" Mike said.

"No point, Mike. I need to know what she knew or knows about that operation."

Mike turned to Alina. "Tell her what you know."

Alina straightened her skirt, shifted in her chair and said, "A lot of girls in Russia, where I come from, told me about Russianbryde. It was on the internet, so I looked at it."

"OK," Sandy said, "go on."

"I put some pictures of myself on it. Nothing bad, just nice pictures my mother took of me. I got a lot of men sending me messages. One of them was Mr. Weber. He seemed very nice and educated. He sent me pictures of his house. After a while, he asked me to marry him. I was shocked. I told my mother. There was a phone number on the site—the Russianbryde site—and my mother called them. Then she told me it was good and that I would be going to the United States. I was excited, and I wasn't worried because I trusted my mother and because a lot of my friends went to the U.S. that way."

"Did you ever hear from any of your friends after they went to the U.S.?" Sandy asked.

Alina took time to think.

"No. Never."

"That's what I thought," Sandy said.

Tom Foster nodded, then turned to Michael.

"Mike, we need Alina to do some work for us," Foster said.

"Work? Why? She's not charged with anything, is she? I mean, this isn't a 'cooperate or go to jail' thing, is it?" Mike said.

"No, Michael, it's a 'save lives thing.'" Foster said.

Vince piped up. "Michael, just listen."

Mike leaned back in his chair and crossed his arms in front of himself.

"There's a little matter of a half-million dollars of bail still held by the finance department up in Ithaca on Alina's case."

"The judge released the bail when she remanded Alina to jail after the indictment," Mike said.

"She did," Foster confirmed, "but it's never been recovered by whoever posted it. Apparently, he or she just walked away. Don't you think that's a little odd, Mike?"

"Well, sure. A half-million dollars. Who walks away from that?"

"That's what we intend to find out, and that's why we need Alina to work with us."

Sandy turned to Mike.

"Michael, we have intelligence that tells us this Russianbryde is bad stuff, really bad stuff. But there are so many layers to it, corporations, limited liability companies, limited partnerships, shell companies, set up in several different states, all stacked up that we can't get to the principals—the people behind it. We also have information that behind Russianbryde is an all-out continuing criminal enterprise, a money-laundering conspiracy that even includes murder."

"And you think Alina is part of it?" Mike asked, leaning forward in his chair, his hands on the edge of the table.

"No, Mike. She's a *victim* of it."

"So what do you want of her?" Vince asked.

"We want her to reach out to Russianbryde and tell them she has the bail money."

"Wait," Vince said. "You just told us Russianbryde is a bunch of killers, and you want to throw Alina into that mess?"

Sandy turned to Tom Foster. "Tell them, Tom," she said.

"Look, I don't want any of you to worry. We'll keep Alina under constant, non-stop surveillance, with armed agents always within a coin toss from her. We'll keep her safe," Tom said.

Alina took Mike by the arm and leaned toward Sandy. "The girls, my friends, my friends that went with Russianbryde. Are they safe?" she asked.

"This is what we need your help with, Alina," Sandy said.

"Then I will do anything. I will do anything!"

"Thank you, Alina," Sandy said. "I promise we will keep you safe."

Sandy turned to Mary Beth.

"Welcome back to the fray, Mary Beth."

Mary Beth turned to Vince, who shook his head. She then turned to Sandy. "Well, I'm here."

"You have a boat," Sandy said.

Mary Beth gave a half-smile. "We do."

"Quite the yacht."

"It's all wood," Mary Beth said.

Vince shrugged his shoulders.

"And it's smelly," Mary Beth said.

"Well, we haven't been there in a while, so..." Vince started.

"We need you to get down there. You and Vince."

"I take it this is my new assignment," Mary Beth said.

"This and more," Sandy said. "You'll be assigned an undercover team of Special Agents, and you'll all get a briefing separate from this meeting tomorrow."

"Jesus!" Vince said. "A team?"

Mary Beth put her hand on his shoulder. "It's alright, Vince."

Sandy reached to the floor by her chair, picked up a small cardboard box, and set it on the table. "This is for you," she said to Mary Beth as she slid it toward her. Mary Beth took the box, opened it and retrieved her FBI badge, 9mm Glock 19 sidearm, and a box of 135 grain 9mm Critical Duty rounds. She leaned over the side of her chair and casually tucked them into her purse. Vince noticed a broad smile across her face.

Sandy turned to Tom, "Tom?"

"Right," he said, turning to Michael, "We'll meet in my office here tomorrow morning at 9:00."

"Sure," Mike said. "Do we need to bring anything or prepare in any way?"

"No. Just be there—you and Alina. I'll take it from there."

"OK, good," Sandy said. "That'll do it for today. I don't need to tell you all this is strictly confidential—a matter of life and death."

They all nodded, and the meeting came to a close.

As they left the conference room, Vince pulled papers folded in half from his inside jacket pocket and approached Alina.

"Alina, give me a minute, please," he said, gently taking her by the elbow and guiding her to a small table in the hallway waiting area outside the conference room. He unfolded the papers, smoothed them over with his palm, handed her his Cross pen, and said, "Here, sign here," he flipped the page, "and here."

"What is this?" She asked.

"Oh, just some formalities," Vince said, winking at Michael.

"OK. I trust you," Alina said as she signed away.

***

The next morning at 9:00, Michael and Alina met with Tom Foster at his office in the FBI quarters of the Binghamton Federal Building.

Alina was impressed with the high ceilings, massive, heavy, dark mahogany desk, the office-packed bookshelves, and the dozen plaques and awards that adorned the walls. Mike had seen it all before. Sitting on Foster's desk was a white envelope and a black attache case.

"Thanks for coming," Foster said.

"Sure," Michael said, as if they had a choice.

Foster picked up the envelope and pulled out a County Finance Department check in the amount of $500,000, payable to Alina Weber. He handed it to Mike.

"It's not real," Foster said.

"OK," Mike said, looking it over, then handing it back to Foster.

"No. I don't want it," Foster said, handing it over to Alina. "We're going to do this in a few steps."

"A few steps?" Alina asked.

"Yes. First, we're going to have you reach out to Russianbryde. We'll use the 'emergency number' they gave you. I want you to tell them you got the bail money back and you have the check. They'll want proof, so you'll email or text them a picture of it."

Foster pulled open his desk drawer, took out an Apple iPhone, and handed it to Alina.

"This is for you to use to call Russianbryde. The phone is monitored by the FBI 24/7. Every call you make or receive will be listened to and recorded. You'll use this phone to call and tell them you have the check."

Alina listened carefully. "And then?"

"They won't be interested in a check from the County Finance department."

"That's for sure," Michael interjected.

"Right," Foster continued. "They'll probably tell you to cash it and bring the cash to them. At least, that's what we're anticipating."

"I see," Alina said.

"Then you'll cash it."

"But it's fake!" Alina said.

Foster slid the attache case around, and opened it facing Mike and Alina. Inside it was a vinyl navy blue zippered six by nine bank deposit pouch.

"What's this?" Alina asked.

"It's a half a million dollars cash. This is what you're going to hand over to them when the time is right."

Both Mike and Alina were surprised at how small the package was. Mike picked it up, unzipped it and looked at its contents: five packs of one hundred, one thousand dollar bills, each just a quarter-inch thick.

"I expected that attache case to be packed with cash," Mike said.

"Oh, we don't do it that way anymore," Foster said, "but I will tell you all the serial numbers have been recorded, and every bill is marked with UV markings."

Mike held the cash in his hand and took a minute to let things sink in.

"So you want Alina to meet with these guys and hand over the cash?"

"Yes," Foster responded.

Mike shook his head. "No."

"I'm sorry, what?" Foster said.

"I said no; we're not doing this. This is putting my client in serious, maybe mortal danger," Mike said as he placed the cash back in the attache case.

Foster snapped the attache case shut, tightened his lips, and leaned back in his chair with his arms crossed in front of him.

"No!" Alina said. "I'll do it!"

"Alina..." Mike started.

She shook her head. "I'm doing this, Michael, and that's final."

"The next meeting will be at your law firm, Michael. You good with that?" Foster asked.

"Of course."

"Good. Let's make it 8:00 p.m. See you tomorrow."

# CHAPTER THIRTY-THREE

Mike filed the New York Criminal Procedure Law section 210.20 motion to dismiss the indictment against Alina. Gary filed a one-page affirmation stating the People of the State of New York have no opposition to the motion. Judge Jennifer Britton called an emergency hearing.

Ken transported Alina from the jail to the courthouse. Mike was already at the defense table waiting for her. He'd called Jason Valor, the social worker, to be there was well. Gary leisurely strolled into the courtroom, in no hurry for anything and knowing full well what was about to transpire. Alina was escorted in her orange jail jumpsuit, handcuffs, ankle shackles and chains—still, at that moment, a murder suspect.

"All rise! The Honorable Jennifer Britton presiding!" The bailiff announced.

All in the courtroom stood.

Britton tucked her robe under her and sat at the bench, immediately reviewing the papers in front of her without looking up at anyone.

"Oh," she said, finally looking up, "be seated. Mr. DeMarco, this is your motion. Proceed."

Michael stood. "Thank you, your honor. I move our 210.20 motion to dismiss the indictment for the reasons set forth in our papers before you."

"You have nothing to add?" The judge asked.

"Well..."

Gary spoke up. "If I may, judge?"

"Go ahead, Mr. Becker."

"Not only do the People not oppose the motion, we *join* the motion to dismiss the indictment."

"Well then," Britton said, "Then I have no choice. The indictment is dismissed, the defendant is released." With that, she struck the gavel, stood, and left the bench, her black robe floating behind her.

Mike turned to Alina. "That's it, Alina, the case is over; you're free." He motioned to Rob the bailiff, pointing to Alina's handcuffs, and the guard promptly removed the cuffs, the shackles and the chain around her waist. "Good job," he said to Mike.

"Thanks, Rob. It was mostly my dad."

"Sure, kid, sure. Well, she gets one more ride in the county limo with the red star on the side. We'll get her up to the jail to get out of the orange and get her stuff."

Just then, Jason Valor came up behind Mike.

"No need, guys. I've got clothes for her right here," he said, pointing to a duffle bag in his left hand. "I don't think she had much of anything at the jail anyway. Can she change here in the ladies room?"

"Sure," Rob said. "That's fine. Just get me the jumpsuit. God knows we don't have enough of them to go around."

To this point, Alina hadn't said a word. She was confused, excited, frightened, and not sure what had just happened. A million emotions bubbled up in her at once. Jason reached for her hand and guided her back through the bar gate toward him. She suddenly launched herself into his arms, went limp, and sobbed. He held her tight while he, Mike and Rob all exchanged glances.

"You got this, Jason?" Mike asked.

"I got it, Mike."

"OK. She's your concern now, for now. Get her something to eat."

Jason gently escorted Alina out of the courtroom; Mike looked at Rob, who shook his head. "Good luck, Mike, and tell your old man I said hi."

"Will do, Rob."

Mike gathered up the papers on the defense table, and left the courtroom, left the building and headed back to his office.

Jason handed Alina the duffle bag and waited outside the ladies' room in the hallway for her to change into the clothes he'd brought for her from the shelter. He waited…and waited, then looked at his watch.

She appeared wearing an aqua-blue and white-flowered bare-shoulder dress, white two-inch heels, with a white cashmere sweater draped over her arm; her long hair brushed, flowing and full across her shoulders. Jason was stunned.

"My God, Alina, you look wonderful."

She gave him a broad, brilliant smile, tossed her arms around him, and gave him a lingering kiss on the cheek.

"Did you buy these?" She asked.

"Oh no no no," he said, gently freeing himself from her arms. "These are all things donated to the shelter."

"Yes! There was a Dior make-up set and hairbrush in there!"

"You'd be surprised how generous people are," Michael said.

She stroked the sweater across her arm. "So I see," she said.

"Come on, let's get something to eat," Jason said, a new cheerfulness in his voice. "Yes! Let's go!"

As they turned the corner into the main hallway of the courthouse, they were greeted by Detective Ken Ludlow and FBI Special Agent Tom Foster. Ken had a dark look about him, something Jason had not seen before.

"Where's Mike?" Ken asked.

"He just left about twenty minutes ago," Jason said.

"Well, better call him," Foster said. "Meanwhile, why don't you just wait in here," he said as he opened the door to the room usually used for grand jury proceedings. The room was empty.

***

Mike was just walking into the front door of the law firm office when Miranda—the phone to her ear—held up an index finger signaling Mike to stop dead in his tracts.

"He just walked in," she said into the phone. She covered the mouthpiece with her palm and whispered, "It's Jason Valor. Looks like the Feds are looking for you."

"Feds? Now what? I'll take it," Mike said, half sprinting to his office.

Mike dashed into his office and snatched up the receiver.

"Jason?"

"Yeah, Mike. Alina and I were just leaving the courthouse, and Ken Ludlow and an FBI agent stopped us and asked where you were."

"Thanks, Jason. Put Ken on the phone for me, please."

"Sure," Jason said, handing his phone over to Ken."

"Ken, what's going on?" Mike said.

"Just get back down here, Mike," Ken said flatly. "Just get down here."

Mike slammed the phone down and bolted out of the office.

"Mike?" Miranda said, as he dashed by her desk, "Are you..."

He was gone.

***

Mike got to the courthouse in twenty minutes, flashed his court ID to the guard, and sprinted up the long marble staircase toward the courtroom.

"They're in the grand jury room!" The guard called out at Mike.

Mike made it up the stairs, two-at-a-time, turned down the hallway and entered the grand jury room.

Tom Foster sat at the head of the twenty-foot-long conference table, Ken to his left, Alina next to Jason to his right.

Foster stood to greet Mike.

"Thanks for getting here so fast, Michael," Foster said, reaching out and shaking Mike's hand. "I've heard a lot of good things about you and your father."

Mike took a breath and tried to calm down. He patted Jason on the shoulder. Alina reached her hand to Mike for a quick shake, and Mike sat down next to her.

"What is this all about?" Mike asked.

Foster reached into his inside jacket pocket and pulled out a document folded in threes.

"I didn't want to serve this on her without you present, Mike," he said as he handed it to Alina.

Mike promptly snatched it from her and unfolded it.

"A Rule 17 subpoena," Mike said as he quickly scanned it. He was relieved to see Alina was not the named target on the document but the witness being summoned, at least as far as he could tell.

The subpoena was issued at the request of the United States Attorney by the Washington District Court of Columbia Federal Court, 333 Constitution Ave. NW, Washington, DC 20001; the title of the proceeding *"The United States of America vs. John Doe 1, John Doe 2, John Doe 3 and Jane Doe.* The subpoena bore the seal of the court. and commanded Alina to attend and testify at the time and place "to be determined." On the second page was a list of items

Alina was to produce to the U. S. Attorney's office.

Mike read it over twice, then shook his head.

"At a time and place to be determined?" he asked.

"Yeah," Foster replied.

"That's not enforceable," Mike said matter-of-factly. "You know that."

"Well..." Foster started.

"Well, nothing," Mike shot back. "Is it for trial? Is it for the federal grand jury? What the heck is it for? You can't subpoena her to your office; you know that. And the U.S. Attorney can't subpoena her to her office either. So what is it? Is she a target or a witness? I sure can't tell," Mike said, sliding the subpoena across the slick table top to Foster.

Foster left it where it stopped.

"You refusing the subpoena? Mike? Really?"

"I am. Really. And John Doe 1, 2 and 3? And Jane Doe 1? How do I know that's not Alina?"

Ken could see this was going badly.

"Mike, call your dad," Ken said in the friendliest voice he could muster.

Mike's eyes bugged. "Call my dad? You kidding?"

"No, Michael, I'm not kidding. Call your dad. Go out into the hallway and call Vince."

The low tone, gravity in his voice, and knowing Ken had always been a friend convinced Mike to swallow his pride and call his father.

"I'll be right back," Mike said as he stood to leave the room. He turned to Alina, "don't say anything while I'm gone."

She nodded; she'd had no intention of saying anything anyway.

Mike called Vince out in the hallway, keeping his voice low as the marble walls and floors would echo his words down the hallway if he wasn't careful.

"John Doe 1, 2 and 3, and Jane Doe 1," Vince said after Mike described the subpoena to him. "Yeah. They can do that, Mike."

"But 'at the time and place to be determined'?"

"I agree. That doesn't comply with Federal Rules of Criminal Procedure Rule 17."

"So, should I tell them to sh..."

"No no no. Take the subpoena, and tell them we'll cooperate one-hundred percent no matter what, and we'll bring her in for a proffer, or reverse proffer, or whatever it takes to comply. OK?"

Mike swallowed hard. "OK, Dad."

"At least it's not a target letter. I had half expected that," Vince said.

"Thank God for that," Mike said.

Mike took a moment to collect himself and returned to the conference room. Foster held the subpoena in his hand.

"I'll accept the subpoena," Mike said. "We'll cooperate completely and voluntarily. Just name the time and place."

"Thank you, Michael," Foster said, standing and offering his hand to Michael. "Assistant U. S. Attorney Sandy Mitchell would like to meet with you and your client early next week."

That's fine," Mike said. Just tell us when and where."

They shook hands, and Foster handed him back the subpoena. "We'll be in touch."

Ken smiled at Mike and gave him a wink. Mike returned it with a knowing nod and tight-lipped smile.

Foster and Ludlow left the room, leaving Michael, Alina and Jason. Mike waited until he was sure they were gone, then turned to Alina. In her blue and white dress and bare shoulders, she looked small and vulnerable. She looked to him with her eyes wide open in anticipation but said nothing.

Mike cleared his throat and looked over the subpoenas again, buying himself time to think about how he was going to explain this to her. He cleared his throat again and noticed Jason was staring at him as well.

"Is it OK? Jason stays?" Alina asked.

"That's up to you, Alina," Mike said.

"He stays," she said, smiling at Jason.

"Good enough," Mike said. "So, here's what's going on," he continued. "This is a subpoena. All it means is that the federal government thinks you have information that could be helpful to them in a case."

"What case? Is my case over?" she asked.

"Your case about your husband is over, Alina. We all know you didn't kill him, so this would be about something else."

Alina looked at Jason, then back to Mike. "What something else?"

"I don't know," Mike said, feeling somewhat either helpless or incompetent. "I really don't know, but we'll set up a meeting with them, and we'll find out. I promise."

"That's it?" Jason asked Mike.

"That's it," Mike said.

"Good! Then, let's go have a great dinner, Alina! We have a lot to celebrate," Jason announced.

She smiled, and the three left the room: Alina and Jason to go eat, Mike to go see Vince.

***

Mary Beth leisurely flipped through the mail she'd just retrieved from the mailbox at the cottage as she walked toward the house—mostly junk mail. Then she stopped dead in her tracks. It was an envelope from *OPM United States Office of Personnel Management*. She quickly tore it open and pulled out the letter inside. Across the top center of the document in gold leaf, it read:

***"Reinstatement of Federal Bureau of Investigation Special Agent***
***Mary Beth Allison."***

*Oh, Sandy,* she thought. *What have you done?*

She walked slowly to the house, the letter in her hand, not looking forward to sharing it with Vince.

# CHAPTER THIRTY-FIVE

**Miami**

Stan Reddinghouse, the Russianbryde tech, was enjoying a dinner of stone crab at the famous Miami "Joe's Stone Crab Shack" with his date when his phone vibrated in the inside pocket of his lime-green jacket. He slid it out, looked at the number calling, turned to his date, and said, "Sorry, I gotta take this."

She frowned and took a long sip of her Cosmopolitan cocktail. He excused himself from the table and went outside onto the deck overlooking Biscayne Bay and answered the call.

"Russianbryde," he answered.

"Ah. Yes. Russianbryde. This is Alina Weber."

Stan stiffened. "Yes?"

"I have a check."

"You have a check?"

"Yes, I have a check, but it is not my money," she said.

"I see. What do you mean?"

"It's the bail money."

He knew exactly what she was talking about. *How the fuck did she get that money?* he thought.

"How much is it?"

"It is five-hundred-thousand dollars."

"I see. And who's the check made out to?"

"It's made out to me," Alina said.

*What the fuck!* he thought. "Are you sure?"

"Yes. I have it right here."

"Send me a picture of it. You can use this phone number."

"Oh. Alright, I'll try. I'm not very good at these things."

The call went dead.

"Fuck!" Stan said aloud. Others having drinks out on the deck turned to him. He waved them off.

Then, a picture appeared on the screen of his phone. It was indeed a County Finance check made out to Alina Weber for a half million dollars, with "bail release" in the memo section.

Then nothing.

He expected her to call back right away, but nothing. He called her number, and it went to a mechanical voicemail.

"Fuck!" he said again, as he dialed Ron Bryce.

"You're never going to believe this," Reddinghouse said.

"Don't make me guess, Stan," Ron Bryce responded.

"That girl, Weber, the one in the murder case up in New York, called. She's got a County Finance check made out to her for half a million dollars. The bail money."

"WHAT? THE FUCKING BAIL? Didn't Holdings get it back?"

"The fuck I know. All I know is that she sent me a picture of the check on my phone. It looks legit."

"I don't understand. How in hell did she end up with it? And since when do they give the bail money to a person that didn't fucking pay it? And where the *fuck* is Holdings?"

"Ron, I don't know, I don't know, and I don't know. How the fuck I know how they do things up in the hicks in New York."

"Well, that check is no good to us. We can't do anything with it."

"I know. But it doesn't seem she intends to keep it. I mean, if she did, she wouldn't have called us."

"That's true. How'd you leave it with her?"

"Not sure. She sent me the picture and never called back. I tried to call her and got voicemail."

"Well, keep trying, and if you have to, get up there, make sure she cashes it, and make sure she doesn't run off with it, and move fast. For all we know, she's already taken off with it or at least deposited it somewhere. She does that; we're fucked."

As he was speaking with Ron Bryce, a second call came in on his phone.

"Hold on, Ron. I'm getting another call. I think it's her. I'll call you back."

He clicked off with Bryce and answered the other call.

"Russianbryde," he said.

"Uh. Yes. This is Alina Weber. Are you the person I was just speaking with?"

"Yes, Alina, it's me. I saw the check, and you're right. It's not your money."

"I know this," she said.

"Right. So, since the check is made out to you, we need your help so we can get it back to the right person, the person who paid it. Understand?"

"I do. I can sign the back of the check and mail it to you or him."

"No no no. I need you to go to the bank and cash the check, then we will come pick it up. Can you do that for us?"

"I don't know."

"Well, listen," he said, frustration creeping into his voice, "the person was nice enough to pay that for you and get you out of jail, so the least you can do is cooperate with us."

"Mr. Volkov," she said.

"Sure, 'Mr. Volkov,'" he said, wondering what in hell happened to Holdings, "Mr. Volkov," and why he never retrieved the bail money.

“I don’t use any bank,” Alina said, “my husband took care of money things.”

“That’s OK, just go to the bank that’s on that check; they’ll cash it, I’m pretty sure.”

“I’ll try.”

“Yeah. Try. Well, better yet. I’ll come up there and take you to the bank. You know, just make sure you’re safe.”

“Oh, that would be nice,” Alina feigned.

“Sure. I can fly up tonight. Where are you staying?”

“I’m at the women’s shelter in Ithaca.”

“Hm. Well, I’ll call you in the morning.”

He ended the call and tapped Bryce’s number. Bryce answered on the first ring.

“I’m going up there,” Reddinghouse said. “Tonight.”

“Good idea. Take the jet. I’ll call our pilot and have it ready for you.”

“OK great.”

“And take Angel with you. You are going to need him.”

“Jesus, Ron, really? He’s a fucking animal. Remember what he did the last time?”

“Sure I do. That’s why nobody fucks with me. Nobody. And when you get the cash, you bring it straight here to me in Annapolis. I’m on my boat. Understood?”

“Understood.”

“Angel” was Ashely Blair. He’d taken so much bullying about his “girly” name, Ashley, as a child his mother nicknamed him her “little angel.” Angel stuck. Bad as it was, ‘Angel’ was better than ‘Ashley.’ The bullying also made him an angry, trigger-tempered, overly-aggressive and violent bully himself.

***

Alina called Mike and told him everything that had transpired on the phone with Russianbryde. Mike, in turn, called FBI Special Agent Tom Foster.

"We'll be ready," Foster said.

"I sure hope so," Mike replied.

***

The fifty-million-dollar Falcon 8X jet, with its two passengers, lifted off Fort Lauderdale's runway 9 an hour and a half later, soaring east out over the Atlantic at over five hundred miles per hour, then turned left for a heading of 340 degrees for a direct two-hour flight to Ithaca.

"You probably shouldn't be drinking so much, Angel. It's a short flight," Reddinghouse said after Angel slammed a fifth shot of tequila.

"Here's an idea," Angel responded, "why don't you shut the fuck up."

"Sure," Reddinghouse said.

"You got your phone with you?" Angel asked.

"Yeah."

"Oh, good. Let me see it."

He handed the phone to Angel, who promptly crushed it in a powerful twisting motion.

"What the fuck?" Reddinghouse said.

"You're a fucking moron, Stan. You really want to be traced on a trip like this?"

"Oh."

"Yeah. 'Oh'. Here," Angel said, handing Reddinghouse a twenty-dollar Tracfone. "For a smart guy, you're really stupid."

"Anyway," Stan said, "we're just going to pick up some cash."

"Is that what you think this is all about? Don't you wonder why *I'm* on this trip?"

“Oh, fuck.”

“Yeah, ‘Oh fuck.’ You really are a dumb son-of-a-bitch. You get the cash; I’ll clean up the mess up there.”

“The girl?”

“I’ll clean up that mess. Why do you think Ron calls me ‘Mr. Clean’?”

“I always thought it was because of your bald head and earring,” Stan said with a smirk.

Angel cuffed him across the side of his head.

“Ow! What the fuck!”

Reddinghouse, Russianbryde’s computer geek, five-foot-ten, one-hundred-thirty pounds, was terrified of Angel, shaved head, six-foot-six, and two-hundred-thirty pounds of solid prison-ripened, tattooed muscle.

Angel was terrified of flying. He took two more shots.

***

**Ithaca**

At the Heights Cafe, Jason Valor sat across the small, round dining table from Alina, their knees almost touching,

“I was surprised you called, Jason,” she said.

“Pleasantly surprised, or just surprised?” he asked.

She flashed him a broad smile, “Pleasantly, always pleasantly.”

“Well, I appreciate you coming, Alina. There’s something we need to talk about.”

“Is it serious?” she asked.

A server interrupted their conversation, placed an empty wine glass in front of each of them and filled the glasses with Merlot. “Are you ready to order?” he asked.

“Oh. We haven’t even looked at the menu yet,” Jason said.

"Take your time," the server said, turning on his heel and walking briskly away.

Jason waited until the server was out of earshot.

"The judge released me as your court-appointed social worker."

She nodded gently. "I see. What does that mean, Jason?"

"Well, it means that I'm not under a court order to look after you like I've been doing."

Her smile vanished. "So, you're done?"

"That's what I want to talk about. I'm not done. I mean, I'm done with the court assignment, but Alina, I don't know if you know, but I'm...I mean I've..."

She reached across the table, took his hand, and their eyes met.

"Me too, Jason, I'm...and I've... also fallen in love... with you."

This startled him; a lump rose in Jason's throat, and he tried to talk but couldn't. He took a deep breath and a sip of wine. A single tear trickled down Alina's cheek. Then, they both began to giggle like two little children.

When they calmed, Jason went on.

"So, Alina, it's not just that my court assignment is over. I've decided to quit as a social worker."

"Why Jason? You've been so wonderful to me, and I never could have survived this without you."

"Because, Alina, I intend to be with you, to love you forever. However, the rules of ethics say that a therapist can't have a romantic relationship with a client until at least two years after therapy ends. I can't wait that long. I don't know what else to do; I would probably lose my license."

"I don't want to wait at all," Alina said. "But Jason, you were *not* my therapist."

"What?"

"No, Jason. You helped me, you got me clothes and food, you gave me transportation, yes, but I know what a therapist is, and what a therapist does, and you were not my therapist; you were my friend, my supporter and my helper."

"Is that how you see it?"

"Yes, Jason. That's how I've always seen you. I was not your client; the *judge* was your client."

"Oh, Alina, that's hilarious, but I like it!"

"Can I ask you a serious question, Jason?"

"Of course."

She paused, taking a long sip of water.

"What is it, Alina?"

"Do I have to stay at the shelter tonight? They've all been so good to me, but..."

"Not if you have a place..., Wait! No, no you don't. Will you stay with me tonight?"

"That's not what I meant, but yes! Yes, I will!"

"You can stay forever, as far as I'm concerned," Jason said.

"You're going to make me cry, Jason."

"Then it's agreed," Jason said. "Let's order food; I'm hungry."

They chinked their wine glasses.

After dinner, they went to the shelter and retrieved what few items Alina had. From there, they went directly to the DeMarco Law Firm office for Alina's meeting with Michael and the four-man FBI team assigned to her. She informed them she was not returning to the shelter but would be staying with Jason at his apartment.

"What do you think of that?" Michael asked Special FBI Agent Manny Lopez, who was in charge of the operation.

"I like that better," Lopez said. "Fewer people exposed to collateral damage."

Mike shook his head and looked over at Jason—the potential "collateral damage."

Jason shrugged.

"Well, here it is," Lopez said, lifting the attached case from the floor next to his chair, removing the blue bank deposit pouch, and handing it to Alina. "A half-million bucks cash. We're going to trust you, Alina."

This struck her funny, and she gave a quick chuckle. "Where would I go?" She said.

"Does Alina call them?" Mike asked.

"Nope. Just wait; they'll reach out to her," Lopez said.

"Manny, here," FBI Special Agent Al Larzare said, handing Lopez two small black leather-covered boxes. Lopez took one of the boxes, set it on Mike's desk and opened it.

"We're going to wire you both," Lopez said. "We'll start with you, Jason."

"Me?" Jason said, slightly alarmed.

"Yes, you, Jason. We expect you'll be in the thick of things when this goes down. You OK with it?"

Jason glanced at Alina, "Of course I am. I'll be right by her side the best I can."

"Agent Mary Ellen here will wire Alina. Mary Ellen, take Alina into the other office and set her up." He handed Mary Ellen the second black box.

"Roger that boss," Mary Ellen said to Lopez. She took the box, took Alina by the hand and led her out of the room.

***

**Annapolis**

The deepest part of the Chesapeake is just southeast of Annapolis, near Bloody Point. It's almost two hundred feet deep and called "The Hole." Like most of the Bay, the downward currents are relentless and powerful. "The Hole" is known to never give up its dead. It was 2:00 a.m., and that's where the two vessels were headed.

John Holdings, now an unwitting passenger in his own 1930 Frank Anderson Yacht bilge, was joined by another unwitting passenger, John Miller. Bill Brown captained the Frank Anderson; Randy Spaulding captained Miller's 1929 75' Elco Yacht, following Brown in tandem. Brown watched carefully as the depth sounder showed the Bay deepening until it registered two hundred feet. He lowered the throttles to idle and shifted the gears into neutral. He went to the bow cabin, pulled up the floor panel, located the thru-hull fitting where water from the sinks onboard all drained into the sea, and with his knife, sliced the two-inch rubber hose attached to it. Water immediately began gushing into the bilge from the Bay like a fire hose.

Spaulding pulled the old Elco Yacht alongside the Frank Anderson, and Brown leaped aboard. The Frank Anderson was already listing to the port side.

"She'll be gone in a few minutes," Brown said.

"Yeah. Sad. What a pity. What a great boat."

The wounded vessel slumped more to the port side; the stern surrendered to the depths, and one minute later, the bow raised up to the sky as if gasping for one last breath, and the boat disappeared into the abyss of the Bay stern-first.

And so, John Holdings, aka "Nikolai Volkov" and John Miller went to the bottom.

"Boy, that was a great boat," Bill said.

"Yeah. What a pity. You hungry?"

"I could eat."

"Me too. Let's get out of here."

As the Frank Anderson made its way to the bottom of the "Hole," the pressure of the deepening water crushed the frail wooden hull, releasing broken planks, cushions and debris to float to the surface. Diesel fuel leaked from the fuel tanks' air vents, making its way to the surface and creating a multi-colored stream flowing down the Chesapeake. By daylight, the slick would be fifteen miles long and a mile wide.

Fifteen-year-old Amy Abrams and her father, John, were die-hard crabbers whose family had made their living on the Chesapeake for three generations. As much as they loved anything or anyone, they loved the Chesapeake. Amy steered the vessel as they made their way upstream in their twenty-foot Boston Whaler boat, headed for their favorite crabbing spot to pull up the traps they'd set the day before. Suddenly, they found themselves cruising through the diesel slick. The sight of the rainbow-colored contamination broke their hearts. As they continued on, they maneuvered to avoid long planks of white-painted wood, life jackets, and a life ring.

"Stop the boat," John said.

Amy pulled the throttle to idle and shifted the engine into neutral. The boat quickly came to a stop. John grabbed the long boat hook, hooked the white life ring as it came alongside, and pulled it onboard.

"Look at this," he said.

"Blue Crab Shack," she said, reading the black letters painted on the life ring. "That's the name of the boat?" She asked.

"Could be. I know it's the name of that restaurant downtown Annapolis."

"Well, it's not a very good name for a boat," she said. She dialed 16 on the VHF radio set and handed her dad the microphone. "Here, Dad, better call the Coast Guard."

"Pan-pan, Pan-pan," John called to the U. S. Coast Guard.

"Pan-Pan" is the internationally recognized radio call that someone aboard a vessel uses to request help at sea when the situation is urgent, but there is no immediate danger to anyone's life or to the vessel itself. It's one step below "Mayday."

"Coast Guard Annapolis," the response came.

"This is the *AMYBELLE*, a twenty-foot fishing vessel."

"Go ahead, *AMYBELLE*."

"We're just drifting about ten miles southeast of Annapolis in what appears to be a debris field from a sunken boat. Lots of diesel in the water, planks, a few lifejackets."

"Roger that, *AMYBELLE*. Any indication of survivors?"

"Not so far."

"OK. We'll send a boat out right away. Please stand by the location."

"Oh, and we retrieved a life ring with the name *'Blue Crab Shack'* on it," John said.

"Roger that," the young officer said, clicking off the call.

He turned to his young colleague sitting at a small metal desk to his right, "Holdings' boat," he said.

"The old Frank Anderson yacht?"

"Looks like."

"No shit," his young colleague said, "no shit."

"Yeah. And how many times have we had to tow that old boat in?"

"I've lost count. Beautiful boat, though. But those old engines were never very reliable."

"Yup, beautiful boat, a real pity."

"Holdings must be devastated."

"Hope he's OK. I like the old guy," the younger one said.

"I'm sure he's fine. Probably took the dinghy in or got picked up by one of the other guys at that marina. Those guys are all really close."

# CHAPTER THIRTY-SIX

## Ithaca

District Attorney Gary Becker was just returning to his office from a lunch of a bacon cheeseburger and French fries that he was already regretting. As he entered the foyer, Detective Ken Ludlow was there, sitting in one of the straight-back old wooden chairs arranged against the wall in the waiting area. He stood as Gary entered.

"Gary, you got a minute?" Ken asked.

"Yeah. Just about a minute. What you got?"

"You're gonna love this," Ken said.

"OK," come on back.

When they got to Gary's office, Ken pulled a red and black flash drive from his shirt pocket. "Wanna watch a movie?"

"Movie?"

"It's short, and I promise you're gonna love it."

"Sure," Gary said as Ken handed the flash drive to him.

Gary slid the drive into the side of his laptop and leaned back in his chair. "Should I make popcorn?" he asked.

"Just watch," Ken said. "There are two separate clips."

The screen lit up with the first clip.

"Whoa!" Gary exclaimed. "I didn't know the old Circles Building had surveillance! That's hilarious."

"Keep watching," Ken said.

The scene was down the hall of the second floor of the Circles Building. The first clip showed a middle-aged woman going into Adams' office, then leaving minutes later just as a massive, red-haired, red-bearded man entered. A few minutes later, it revealed Sean Adams, apparently drenched in some dark liquid, bolt out of his office, and run down the hallway to the stairs. Minutes later, the clip showed the massive man in a hoodie leave in a rush, sprint down the hallway and exit the back door.

"Wow," Gary said.

"Yeah, wow," Ken echoed. "This was what was going on when you and I encountered Adams outside the courthouse, all drenched in coffee. Play the next clip."

The next scene was from security footage of the front of the hospital. It showed various cars pulling up, people getting in or out of them.

"OK," Ken said, "here it comes, watch."

The clip showed Adams hiding behind one of the massive pillars at the entrance of the hospital, dressed in nothing but a hospital gown and yellow socks, then a large black SUV stopping at the entrance. Adams scurried to the vehicle, tapped on the window, and got into the back seat. The SUV took off quickly.

"Play it again, Gary."

He played it again.

"Now focus on the passenger side window when it rolls down."

"Ah," Gary said. "I see red hair and a red beard. That's our guy, 'Big Red.'"

"There's more," Ken said. "I've got the autopsy report on Adams' death." He pulled it out of his jacket inside pocket, unfolded it and handed it to Gary.

Gary read it aloud, “Cause of death: Congestion of the head, cyanosis, petechial hemorrhages, bleeding from the mouth, nose and ears, compounded by hypothermia. Homicide.”

He set it down on his desk. “Homicide strangled. The guy fucking murdered Adams.”

“Looks that way.”

“Did you get an ID on the vehicle? Plate number?”

“Sure did. Camera caught it just as it left the lot.”

“And?”

“Titled, registered and insured in the name James Curtis, aka our ‘Big Red.’”

“You gotta be kidding!” Gary said with a chuckle. “He’s all legit with his car?”

“Hey, nobody said he was a genius, Gary.”

“And a black Land Rover,” Gary said. “Not too many of those around.”

They both took a minute to let it all sink in.

Gary piped up, “Maybe Big Red provided the gun to Adams for the purpose of killing Weber for Alina. Three-way conspiracy. Then Big Red has to kill Adams to cover his ass.”

“Makes sense, Gary. Makes sense. Adams was the link between Red and Alina.”

“I knew there was more to this Alina Weber than everybody thought. I knew it. Well, we gotta get this guy in fast, at least for questioning as a possible murder suspect. We don’t have enough to arrest him, just speculation.”

“Shouldn’t take long, I notified the state police and his plate number is already up on the Automatic Number Plate Recognition system. He’s bound to use a major highway at some point. Either way, there’s an all-point bulletin out for his SUV. It won’t take long.”

“Make sure it’s an ‘armed and highly dangerous’ APB.”

"It is," Ken assured Gary. "Should we put the FBI on notice?"

"No. Not yet. I want this one, and I don't want Sandy Mitchell swooping in and taking it."

"Isn't there double jeopardy protection for Alina Weber?"

"No. Not at all," Gary said. "That indictment was for murder in the second degree and we never prosecuted that charge. What we might be looking at now is conspiracy to commit murder. Different crime, same prison time."

"Roger that."

***

Stan Reddinghouse and Angel landed at the Ithaca airport around midnight and went straight to the car rental counters. Only one was open; "Vans R Us."

Zeke Monroe, A very tired and bored clerk, was sitting on a low stool behind the counter. Reddinghouse cleared his throat twice, and the clerk slowly stood up.

"Can I help you?" Zeke asked.

"Yes. We need to rent a car."

"Sure," Zeke said. "Nine or twelve-passenger?"

"What?"

"I said, nine or twelve passenger?"

"No, just two of us. Him and me," Stan said, pointing to Angel.

"I see," Zeke said. "So it will be just the nine-passenger."

Angel approached the counter. "What the fuck is going on?"

Stan turned to him, "Well, it seems there are only vans to rent here tonight. Vans."

"Vans?" Angel asked.

"Yeah. Fucking vans."

"They got nothing else?"

Stan turned to Zeke. "You got nothing else?"

"Nope. Sorrrrry."

At this, Angel decided to take matters into his own hands.

"What are you driving?" He asked Zeke.

"Uh, Subaru wagon, why?"

"We'll take it."

Stan clenched his teeth; he knew what was coming next.

Angel reached into his pants pocket, pulled out a fist-sized wad of cash, peeled off ten one-hundred dollar bills, slapped them on the counter, and said, "Keys."

Zeke took a step backward and said, "I'm sorry, I don't think you understand but..."

*Oh, big mistake,* Stan thought to himself, *HUGE mistake.*

Angel picked up the cash, walked behind the counter, took Zeke's hand, mashed the money into his palm and said, "Keys."

"It's my wife's car..."

"Then buy her something nice," Angel said, "keys."

"Right. Keys," Zeke said, reaching into his right front pants pocket. "Here."

"Good," Angel said. "We only need it for a day or two," he turned to Stan, "Van? Are you fucking kidding? You are such a wimp. Let's get the fuck out of here."

Zeke waited until they'd left the building, then dialed his wife.

"Honey, you're not going to believe this..."

***

Mike was already at the office at seven the next morning, filled with anxiety at what he expected to happen that day. To try to calm his nerves, he went over to Miranda's desk, picked up the stack of mail that had come in yesterday, and flipped through it as he walked to his office and sat at his desk. Among the usual letters from other law firms, court notices, and junk mail was an envelope addressed to Alina Weber c/o Vince DeMarco, Attorney from the Presidential Life and Casualty Insurance Company. He dialed Vince.

"Hey, Dad."

"Good morning, son. What's up?"

"I'm at the office, and there's an envelope addressed to Alina care of you from the Presidential Life and Casualty Insurance Company."

"Wait. Did you say you're at the office?"

"Well, yeah. I'm at the..."

"Why so early?"

"Couldn't sleep."

"I understand that," Vince said. "Go ahead and open it."

Mike took the brass letter opener from his desk drawer and slit the envelope open. There was a pause, and Mike started laughing, and laughing, and laughing...

"Mike? You OK?"

"Oh God, Oh God. Dad! Did you do this?"

"Do what?"

"I'm holding a certified check payable to Alina for five fucking million dollars!"

"Wow, that was fast!" Vince said.

"So you *did* to it!"

"Yup. Remember I had Alina sign a couple documents at the courthouse?"

"Vaguely."

"That was the claim form for Weber's life insurance policy."

"But that was just a few days ago."

"Michael. Don't you know me better? I've been working with the underwriters and field claim adjuster ever since Adams told us about the policy. They were ready and expecting the final claim form. I filed it with them electronically, and they had already set up the reserve and were ready to issue the check when they got those last two documents. It didn't take much: a copy of the court disposition dismissing the indictment, Weber's death certificate, a quick petition for letters of administration naming me the administrator of the estate and a sole beneficiary affidavit."

They both paused to think.

"Dad."

"Yes?"

"Is it possible that Sean Adams wasn't as stupid as we thought?"

"What do you mean, Mike?"

"I mean, we thought he was an imbecile because he told us Weber's ex-wife would get that money, and that would exonerate Alina from the murder charge."

"Yeah, I remember."

"What if he was smarter than that? He was a Cornell grad, after all."

"I don't get it."

"I've got a check for five million dollars payable to Alina Weber on my desk. Supposing, just supposing, Sean Adams knew damn well Helen Stinson Weber wouldn't get a dime of it."

"OK, so?"

"So he killed Weber, unfortunately accidentally let Alina take the fall for it, but knew all along that Helen Weber would be disqualified from the beneficiary clause because they were divorced, that Weber had no children, the life insurance benefit would then have to go into the estate, and Alina would be the only person to get it."

"OK."

"The only thing that I don't understand," Mike said, "was how, or would he have known about the life insurance policy before the murder."

"Mike. He had to know. He was defending Weber from a malpractice lawsuit that exceeded his malpractice insurance coverage. He was sued for ten million; he only had one million in malpractice insurance coverage. Adams would've taken inventory of all Weber's possible assets, including any life insurance policies."

"You already knew all this, didn't you, Dad?"

"Yupper."

They took a moment to think.

"Naw, I'm sticking with the 'imbecile theory,'" Mike said.

"Yeah, me too."

"Look, let's not tell Alina about this just yet. I don't want her distracted while she's in the middle of this thing with Russianbryde," Mike said.

"I agree. Just put it in the office safe for now."

"It won't earn any interest in there," Mike said.

"We'll get it to her quick enough."

# CHAPTER THIRTY-SEVEN

Jason and Alina spent most of the night talking, cuddling and making love. The tension over what was about to go down made all of it more intense. Unable to sleep past six, Alina was relaxing on the couch wearing one of Jason's T-shirts; Jason was in the kitchenette in his bathrobe, pouring two cups of coffee when his cell phone sounded. It was FBI agent Manny Lopez. Jason took a deep breath and answered.

"Hi, Jason. Lopez here. I called your phone not to alarm Alina."

"That's OK. What's happening?"

"We expect Alina will get a call probably around nine or so. Two targets arrived by private jet last night from Fort Lauderdale, and they'll probably go to the shelter looking for Alina. When she's not there, they'll call her. Be ready."

"I'll tell her."

Lopez clicked off the call.

Jason brought Alina her coffee and sat close next to her. "That was Lopez; he said you should expect a call from Russianbryde around nine."

She nodded as she sipped her coffee, holding the cup in both hands.

"Let's get dressed," she said.

***

The 1989 Subaru Forester station wagon pulled up to the curb in front of the women's shelter building. Angel started to get out.

"Whoa, whoa, whoa!" Stan Reddinghouse said. "You're not going in there. You'll scare the shit out of everybody. You stay put; I'll go in."

"Fuck," Angel said, sitting back in his seat.

Reddinghouse exited the car and made his way to the front door of the shelter. There was a security camera over the solid wooden door and a speaker box with a button to the left of the door. He pressed the button.

"Can I help you?" Rachael's voice came over the intercom.

"Yes. Thank you. I'm here to see Alina Weber."

Rachael had been expecting this. No one was to know who was at the shelter. That was the very essence of the protection it provided to domestic violence victims.

"I'm sorry," I can't help you," she said.

"Well, I have something for her that she requested. Can I leave it with you?"

Rachael was well familiar with this ploy, often used by abusers; it was an attempt to determine if their victim was in the shelter.

"I'm sorry, I can't help you," she insisted.

Reddinghouse was counting his blessings that he'd not let Angel approach the shelter. He stood for a couple minutes; Rachael was observing him on the security screen at her desk. He finally turned and walked back to the Subaru.

"Give me that Tracfone, Angel. I gotta call her."

Rachael called Manny Lopez. "They were just here," she said.

"Got it," Lopez responded. "Thank you, Rachael."

"I'm worried," Rachael said.

"It'll all be OK, I promise," Lopez assured her.

Sure enough, Alina's phone, lying on the coffee table in front of the couch, sounded at 9:00. She turned to Jason, her eyes wide.

"It's OK, Alina. Answer it, and remember what Agent Lopez told you."

She reluctantly picked up the phone and tapped "accept."

"Hello Alina?" Reddinghouse said.

"Yes. Who is this?" She asked, already knowing.

"This is Russianbryde. We need to meet today. Do you have the check we talked about?"

She paused, feeling panic rise in her chest.

"Do you have the check we talked about yesterday?" He repeated.

"No."

"No?"

"What do you mean, 'no'?"

"I cashed it like you said."

Reddinghouse put his hand over the receiver, turned to Angel, and said, "fuck man, she cashed it."

Angel gave him a 'thumbs-up.'

"Well, that's good work, Alina. We are all very grateful to you."

Angel snickered under his breath.

"Are you at the shelter?"

"No."

"Oh, OK. Where are you?"

"I'm at a friend's house."

"I see. Can we come over?"

She turned to Jason; he nodded.

"OK."

She gave him the address.

"We'll be there in a few minutes, and thank you, Alina."

"Of course," she said, tapping off the call.

Jason's phone sounded immediately; it was Lopez.

"We're tracking that phone. It's coming from right in front of the shelter, and it's leaving right now. We're all in position, Jason. You ready?"

"I am."

"Is Alina?"

Jason glanced over at her. She sat on the couch upright, her legs crossed at the knee, her arm folded in front of her, wearing tight dark blue jeans, a button-down white shirt, and gleaming white Adidas sneakers. He noticed a stern, determined look on her face he'd not seen before.

"She's ready," he said.

Fifteen minutes later, Stan Reddinghouse and Angel came stomping up the exterior metal stairs to the second floor of Jason's apartment building. Stan knocked on the door. Alina peered through the peephole, and Stan gave her a friendly smile. She opened the door. As soon as she did, Angel burst into the apartment, Stan following tight behind him.

Angel stood off to the side, and Stan wasted no time.

"You have the cash?" he said.

Jason was immediately alarmed. No one expected the likes of Angel to come barging in, and the tone of Stan's voice set off a million red flags in his mind.

"Who are you guys?" Jason asked.

Stan glanced at Angel.

"Who are we? We're from Russianbryde. Your girlfriend here knows all about it."

"No, I mean, who are *you?*" Jason insisted. "We're not going to hand over anything until we know you're really from Russianbryde and that you paid the bail for Alina."

Stan was amused at this young man's gall. He smiled, shook his head, and sat in the gray faux-suede chair across from the couch.

Angel stood behind him.

"Have a seat," Stan said to Jason, motioning to the couch; "you wanna talk, we'll talk."

Jason sat next to Alina.

"Yeah. We're from Russianbryde. We got hundreds of girls like your girlfriend there, shipped all over the country. We don't know where the fuck they end up; we don't care."

Angel pressed his hand on Stan's shoulder, "You really gonna do this?" he asked Stan.

"Sure. What difference does it make? They wanna know, I'll tell 'em."

Stan went on, "It's just that every once in a while, shit goes bad, like with Alina here. Somebody gets killed. Only usually, it's one of the girls. Then we gotta clean it up, keep the business going."

The more Stan went on, the more Jason knew Stan and Angel weren't going to let Alina or him out of this room alive.

"So then we find out Alina killed one of our clients, and our boss down in Miami tells us to get up here and clean it up."

"Mr. Volkov," Alina said.

Stan and Angel laughed.

"Yeah, Mr. Volkov," Stan repeated. "Well, enough of this chit-chat. Where's the cash, and you better fucking really have it."

"I'll get it," Alina said. She stood and left for the bedroom, returning with the cash in her hand. She sat back on the couch and placed it in her lap.

"More chit-chat," she said.

"What?" Reddinghouse said.

"I said, more chit-chat. I want to know where the three girls from my town in Russia are," she demanded.

Stan turned around to Angel, "You believe this shit?" he asked.

"A lotta balls, this one," Angel said.

"Sure," Stan said. "What town?"

"Tolyatti."

"Ah, Tolyatti!" Stan said, "You'll never hear from them again."

At this, all the fear, depression, anxiety, isolation, abuse, and anger stored in her for the past year boiled over. She flung the pouch in Reddinghouse's face and lunged across the coffee table at him. Both Jason and Angel were at first paralyzed in shock, but in what seemed like an eternity, Jason yelled, "NO ALINA!"

She took Reddinghouse to the floor as the chair flew backwards to the floor, and Angel grabbed her by the neck and lifted her off Reddinghouse and off the floor. Jason lunged at Angel and gave him a left hook punch worthy of an MMA fighter. Angel laughed and dropped Alina to the floor.

Angel raised his right fist to take Jason out when the door to the apartment burst open. Manny Lopez teased Angel in the back of his neck, and Angel dropped like a ton of bricks to the floor in spasms. Lopez teased him again, and FBI agent Mary Ellen cuffed him behind his back.

"WHAT TOOK YOU SO LONG?!" Jason screamed, helping Alina to her feet.

"We didn't expect Alina to go on the attack," Mary Ellen said.

Lopez cuffed Reddinghouse, who was bleeding from the side of his head.

Two other male FBI agents came into the room.

"Take this one out of here," Lopez said, pointing to Angel. "Leave the other one here. We're going to have a little talk." Lopez lifted the chair from the floor and shoved Reddinghouse into it.

"I'm bleeding! I'm fucking bleeding!" Reddinghouse said as blood trickled down his face and onto his shirt.

"You sure are," Lopez said. He turned to Alina. "Are you alright, Alina?"

"I haven't felt this good in a long time," she said.

"Jason? You don't look so good," Lopez said. "Better get a glass of water."

Alina took Jason by the hand and led him to the kitchenette.

Lopez sat on the coffee table facing Reddinghouse.

"OK, Stan," he said, "You got a couple options here. You listening?"

"Yes."

"OK good. So, one option is to follow your buddy there to the jail."

"And?"

"The other option is to be smart enough to know the position you're in. From what my intelligence tells me, you're a pretty smart guy."

"I don't understand."

"Oh, Stan Stan Stan, of course you do. You're right in the center of a RICO case. You know what that means, Stan? It means that you're guilty of every crime any of your associates committed: every murder, every extortion, every kidnapping, assault, gun charge, money laundering, wire fraud. You understand that?"

"I knew this was coming!" Stan said, "I fucking knew it!"

"Oh good, you understand."

"Now, I'll give you two more choices. You can lawyer-up, or you can cooperate. You know you have every right to get a lawyer. In fact, you know what? Mary Ellen read him his Miranda rights."

"Sure, Manny." She turned to Reddinghouse. "You have the right to remain silent; Anything you say can and will be used against you in a court of law; You have a right to an attorney; If you cannot afford an attorney, one will be appointed for you. Do you understand those rights, Stan?"

"I do."

"Ah good!" Lopez said. "So?"

"I'll cooperate. I'll tell you anything you want to know."

"OK, take him away," Lopez said, "Oh, and have somebody take a look at his head; he might need a stitch or two."

# CHAPTER THIRTY-EIGHT

## Annapolis

The Friday night Marina Committee Meeting was about to begin. Missing were John Holdings and John Miller. Missing from the marina was the 1930 Frank Anderson Yacht.

Chairman Bill Brown called the meeting to order.

"Gentlemen, welcome to our usual Friday night committee meeting. As you may have surmised, we have a very special guest with us here tonight. Gentlemen, just in case you didn't know him..."

Chuckles around the room

"Captain Ron Bryce, Jr., captain of the magnificent yacht, *EMANCIPATION!"*

They all rose to their feet and gave him an exaggeratedly long-standing ovation. Ron gave a half-ass salute without getting up.

Bill went on. "And since John Holdings is attending to business elsewhere, we won't have his usual wonderful buffet from the Blue Crab Shack. We do have, however, a beautifully prepared buffet for you following what I hope will be a brief meeting tonight from the Red Lobster."

"The Red Lobster? Are you fucking kidding me?" Bryce said under his breath.

"OK, first order of business," Brown said, looking down at a pad on his lap.

Bryce interrupted. "Here's the first order of business," Bryce said. "Where the fuck is my mechanic?"

Brown waited for a response from anyone; nothing came, so he spoke up. "Does anyone know where Rick Anders is?"

"I don't know exactly where he is, but I'll bet he's on another one of his binges," Spaulding said.

Nods around the room and a little muttering.

Bryce smiled. "I see. Very well. Then, let me say this. You fuckers find him by tomorrow night, or I'm shutting the fucking place down. Next order of business, please."

The room fell silent.

Finally, Mel Geeves, one of the newest members of the committee, having moved his 1934 sixty-eight-foot Trumpy yacht there from Baltimore harbor just months ago, raised his hand.

"Yes, Mel," Brown said.

"Ahem, Uh. Thank you, Bill. Second order of business?"

"Go ahead, Mel."

"Thanks, Bill. So, I filled my fuel tanks here at the marina pump for the first time, and I was kinda shocked at the cost."

Chuckles around the room.

"Well, Mel, that's because we don't use regular marine diesel in our boats here."

Nods around the room.

Bill continued. "We only use Type 2 diesel, also known as JetA. It's five times the cost per gallon, but it'll make your engines run smoother, cleaner and give you about ten more horsepower and make them last longer. It's the most highly purified diesel fuel there is. Now, isn't your Trumpy worth it?"

"Isn't that what airliners burn?"

"Sure is. They wouldn't use anything but the best fuel, right?"

"Of course, Bill. That makes sense."

"Ok Good."

"One more thing, if I could," Mel asked.

"Sure, Mel, go ahead."

"That Matthews yacht, *MY CHOICE,* looks to be sitting a bit low in the water and is listing to the port. I've been kinda, keeping an eye on her, and I haven't seen the bilge pumps go off at all. Is anybody using that boat? I mean, it's a wooden boat, and we all know they leak no matter what you do, so those pumps should go off once in a while."

*Fuck,* Bill thought. *What a very bad time to bring this up.* "No worries, Mel. I'll look into it. Next order of business, anyone?" He asked, trying to get away from the subject.

"Wait a minute!" Bryce said. "that's the boat docked right next to my yacht, isn't it?"

Mel nodded.

"Bill, I don't want that fucking thing anywhere next to my boat, you understand? Get that thing pumped out and out of this fucking marina!"

"Of Course, Ron. Of course."

Bill couldn't remember if there was more water, leaking diesel or blood in the bilge of that boat by now. He knew Rick had set it up to blow with a pin-hole leak in the fuel tank under the aft cabin and that somewhere, there was the remote detonator. He looked down at the notes he'd made for the meeting, and the next order of business was about the garbage dumpster being over-filled. He chose to end the meeting. Enough was enough.

"OK, folks," Bill said. "That'll do it for the meeting tonight. Please go and enjoy the buffet."

They all stood to go to eat. Ron Bryce walked directly to Bill.

"I don't like the way this place is being run, Bill. You gotta get your act together. I told you to get that damn Matthews death boat out of here almost a year ago."

"I know, Ron; I know..."

Just then, Bill saw Vince and Mary Beth join the buffet line.

*You gotta be fucking kidding,* he thought to himself. *Now? And where the FUCK is Rick?*

"Who are they?" Ron asked Bill.

"Well, uh..."

"Who are they?"

Bill swallowed hard. "Those are the owners of the Matthews."

"I said, who *are* they?" Ron repeated.

"Well, Ron. That's the lawyer and..."

"The FBI agent? Is that what you're telling me? The fucking FBI agent!"

"That's right, Ron. That's who they are."

Ron put his face within one inch of Bill's face and said in a slow, low, subdued voice, "Here's what you're going to do, Bill. You're going to walk me over there, introduce me to them, and I'm going to invite them for a little visit on my yacht. Understood?"

"Sure, Ron. OK."

Bill and Ron both put on smiles and headed over to Vince and Mary Beth. Vince had a plate full of lobster, corn, and shrimp; Mary Beth's plate was still empty as she looked over the food.

"Well, hello, folks!" Bill broke in. "I notice you came in late. I want you to meet someone."

"Good to see you, Bill," Vince said.

Mary Beth gave them a gratuitous smile.

"This is Ron Bryce. He owns the *EMANCIPATION* docked right next to your Matthews."

"Wow," Vince said. "We saw it from the parking lot when we pulled in. Isn't that the old Presidential yacht?"

"Yes," Ron said flatly. "So look, when you're done eating, I'd love to give you a tour of the boat; give you a little history here."

Vince and Mary Beth exchanged smiles.

"We'd love that!" Mary Beth said. "Vince, are you still eating that?" She asked.

"Well, uh, no, I guess I'm full."

"Great," Ron said. "Let's go. I'll eat after."

***

The water where *EMANCIPATION* was docked was twelve feet deep. Just thirty feet farther out, the bottom sloped sharply down to a depth of forty to fifty feet. When Rick fell between the *EMANCIPATION* and the dock the night the boat came in, he was swiftly swept under the boat, forced down the sloping bottom and lodged under the branches of a water-logged tree that had been there for decades. The remote detonator was still in his front jeans pocket. The salt water had not yet penetrated into the device's plastic case.

***

Ron led the way down the pier toward his yacht. They approached the Matthews.

"That your boat?" Ron asked, pointing at *MY CHOICE.*

"Sure is," Vince said as they walked by her.

"Well, she doesn't look so good. Maybe you better take better care of her," Ron said.

"Right," Vince said, "We don't get down here enough."

Ron sneered. "Well, here she is, the *EMANCIPATION,* my pride and joy," he said, pointing to the gangway to the yacht. At the top of the gangway stairs stood a young man, six feet tall, broad shoulders, sandy blonde hair, dressed in pristine, sharply-creased white slacks, a button-down, high-collar white shirt with gold and blue-striped epaulets on both shoulders and a navy blue bowtie.

Mary Beth almost choked when she saw him and elbowed Vince.

"What?" Vince said.

She motioned to the young man at the top of the stairs with her eyes.

Vince jolted.

"Come on board. Ladies first," Ron said.

Mary Beth made it onto the deck and was greeted by the young man.

"Welcome aboard," the young man said, making a gesture toward the door behind him that led to the yacht's main salon. "And to you, sir," he said to Vince as Vince came aboard after Mary Beth.

"Ah," Ron said, noticing the attention the young man was receiving from Mary Beth and Vince," this is my newest crew member, Erik Karlsson. My first mate got called away—family business, something about his kid getting in trouble. I was lucky enough to hire Erik here in Annapolis; can always use an experienced hand on board—former coast guard."

"A great pleasure to meet you both," Erik said. "This way, please."

The three entered the salon, and Ron started the tour. He pointed out all the original handcrafted woodwork, stained glass windows, paintings by Pablo Picasso, Vincent van Gogh and Claude Monet.

"These are all originals?" Vince asked.

"I wouldn't have them otherwise," Ron said.

Mary Beth was impressed by the Le Creuset cookware in the Roland GmbH stainless steel galley.

"We'd really like to hear about the history of this yacht," Mary Beth said.

"Of course," Ron said, "let's head up to the back deck. I believe Erik has set out my favorite cognac, if you're interested."

"I am," Vince said without hesitation.

They convened on the back deck, each sitting in the black leather and dark mahogany Campeche chairs. Erik appeared from below with three gold-trimmed St. Louis Thistle Gold Brandy Snifters and a bottle of Remi Martin, Louis XIII cognac, on a silver tray. He offered Mary Beth a glass, which she declined, then served Vince, then Ron.

Mary Beth waited until Erik was out of sight, then started, "What an amazing yacht, but before we talk about her history, we have some other history to talk to you about, right, Vince?"

"We do, Vince said. I'm going to mention a few names—you know, for the sake of conversation—and ask you a couple questions about them. You OK with that?"

Ron leaned back in his chair, took a sip of his cognac, and said, "Sure, *counselor,* why not," knowing full well that neither Vince nor Mary Beth would live to tell anyone about anything after this.

"Good. Let's start," Vince said. "Victor Bennington."

"Sure," Ron said. "You bought his boat."

"OK," Vince went on, "Virginia Bennington."

"Oh tragic. Had to die. Horrible accident."

"A lot of 'horrible accidents' around here," Mary Beth said.

"Yeah. We got a lot of widowers here. Sad," Bryce said with a half-smile.

"Randolph Everts," Vince continued.

"Ah yes, Randy. Was once a decent lawyer, just a drunk now."

"Does a lot of estate work for you here, it seems," Mary Beth said.

"People die, people with boats, people with wives. People die. You should know that well enough, *Special Agent* Mary Beth Allison," Ron said with a smile. "Go on, *counselor,* this is actually fun."

"Sure. John Holdings"

"Hm. John Holdings. Doesn't ring a bell," Ron said.

"Let's try this," Vince said, "1930 Frank Anderson Yacht—Blue Crab Shack."

"Hm. Still nothing," Ron said.

"Well then, let's try this," Mary Beth said, "Nikolai Volkov.."

The smirk left Bryce's face. "Boy, you really have been digging, *Special Agent* Allison."

She went on, "Johnny Stevenson, MaryAnn Calais, Naval Cadet Randy Emerson, the sinking and murder on the Potomac River. Nikolai Volkov, Russianbryde."

What Bryce did not know was that no one had yet been able to connect Bryce or Russianbryde to the deaths of Johnny Stevenson, MaryAnn Calais, Naval Cadet Randy Emerson on the Potomac yet.

"Wow, *Agent Allison,* why bother to ask me when you already know the answer? The whole 'Volkov' thing was our way of cleaning up the messes that could've put my business, Russianbryde, out of business. 'Volkov.' Nice common Russian name, like 'John Smith' is here. I sent him wherever a mess had to be cleaned up."

"Like the Weber murder," Vince said.

"Ha! Yeah. Like the Weber murder. And by the way, I want that retainer back," Bryce said. "You didn't earn a dime of it."

"It's right here," Vince said, pulling a white sealed envelope from his inside jacket pocket and tossing it on the oval marble coffee table between him and Bryce. Bryce picked it up, opened it, counted it, and set it back on the table.

"You know, Ron," Mary Beth said in a coy voice, "you're actually quite the genius."

"Don't play me, Allison," Ron said, sitting up in his chair.

"Oh, I'm not playing you, Ron," she continued, "Innocent, naive young Russian girls, desperate men with lots of money, and all you have to do is sit back and reel in the cash. Probably by the millions."

Bryce was smiling now.

"Just a dating site," Bryce said.

"Of course, Ron. Just a dating site that just needs a murder and cover-up once in a while to keep the cash flowing. I can understand that," Mary Beth said.

"What I can't understand," Vince piped in, "is why the Victoria Bennington murder."

"Are you really that stupid, DeMarco?"

"Maybe," Vince said.

"How many fucking women, not counting the FBI agent you're sitting next to right now, have you ever seen in this fucking marina?"

"None."

"Brilliant," Bryce said. "And it's all a matter of *choice.* What's the name of your boat, stupid?"

"*MY CHOICE,"* Vince said.

"Right. Victor made his choice. We've all made that choice here."

"Then why did you have Victor murdered?" Vince asked.

"Oh, you're fishing now, aren't you?"

"No. Why?"

"Simple. He had 'buyer's remorse.'"

"I see," Vince said.

"You know what? This is getting boring, and I haven't even eaten yet. Guys! Come on in," Bryce said.

The door from the salon to the back deck opened, and Bill Brown and Randy Spaulding came on deck. Brown had his Beretta pistol in hand. Spaulding had a handful of black zip ties.

"We're done here," Bryce said to Brown. "They're all yours. Clean this shit up."

"The Matthews?" Brown asked.

"For now," Bryce said. "I gotta get something to eat. Then I want that fucking boat in the Hole, with them on it."

Just then, the new first mate, Erik Karlsson, appeared in the doorway.

"That'll be enough, gentlemen," Karlsson said, brandishing his FBI-issued Glock 19 in one hand and his FBI badge in the other. "FBI Special Agent Tom Foster. Put the gun down, Brown. You're too old to be doing this shit."

Brown dropped the gun.

"Boy, we're glad to see you here, Tom," Mary Beth said.

Bryce pulled his Baby Browning A .25 ACP Pocket Pistol and let off a shot toward Foster, missing him and shattering the stained-glass window in the salon door behind him. Vince lunged at Bryce, shoving him to the floor, the pistol skittering across the deck. Two other FBI agents quickly appeared from below, cuffed Bryce, Brown and Spaulding.

"Wow, Vince! Not bad for a semi-retired trial lawyer!" Foster said.

"Thanks, Tom. And I've never been happier to see an FBI Agent than I was when we first boarded this old boat. Well, except for Special FBI Agent Mary Beth, of course."

"Good save, Vince," Mary Beth said.

"Christ, Vince, this is the *second* time the FBI has saved your life! Either you're doing something very wrong or very right," Foster said. "Come on, let's get out of here."

***

The Chesapeake salt water finally penetrated the seam of the plastic case of the remote detonator in Rick Ander's jeans pocket, shorting out the circuit and sending the signal to the detonator attached to the fuel tank in *MY CHOICE*'s bilge.

The explosion that ensued was nothing short of cataclysmic and instantaneous, with all the type 2 diesel Jet A fuel vapors that had accumulated over several days. The fifty-ton boat leaped into the air twenty feet, toppling the corrugated steel roof from its support beams; a monstrous ball of roiling fire unfurled with unbridled fury a thousand feet in every direction. Nothing in its path was spared, setting everything in its path ablaze, and the boats in the marina exploded one after the other in sequence as the heat burst open their fuel tanks. By far, the largest, most violent, earth-shattering explosion happened when the blast hit *EMANCIPATION.* The yacht exploded with a deafening roar and such force it was felt thirty-two miles away in Washington D.C. Alarms blared across the city, plunging law enforcement into emergency response mode, as they believed the epicenter of the explosion to be within the Capitol itself. The president of the United States was rushed into the basement bunker at the White House, and Congress was ushered into the Capitol Building basement.

Tom Foster, Vince, Mary Beth, Ron Bryce, Bill Brown, Randy Spaulding, and the two other FBI agents had just made it to the end of the pier when the explosion happened. They were blasted into the parking lot and found themselves pinned beneath a thirty-foot by forty-foot sheet of corrugated steel from the marina cover roof. The first to rouse was Foster. He was closest to Vince and couldn't see the others.

"Vince! Vince!" Foster called.

"I'm OK," Vince said. "Mary Beth is right here next to me; I think she's OK."

"I'm alright," Mary Beth responded.

Bryce, Brown and Spaulding were knocked out cold since they were handcuffed when the blow happened and did face-plants into the ground. The other FBI agents were already busy using their considerable strength to lift the steel off the others using their backs against it. Sirens were blaring, and two firetrucks and an ambulance pulled into the parking lot.

Bryce came to, "MY BOAT! MY BOAT! WHAT HAPPENED TO MY BOAT?" he screamed. This woke Brown up.

"SHUT THE FUCK UP, YOU ASSHOLE!" Brown yelled. "I'M GOING TO KILL YOU!"

Foster, Vince, Mary Beth and the other FBI agents managed to lift the sheet of steel just high enough for everyone to crawl.

"COME ON EVERYBODY! MOVE! MOVE! MOVE!" Foster called out, "The whole place is on fire! We gotta get out NOW! MOVE!"

One of the fire trucks was already spraying water on the corrugated steel to cool it and stop the fire from reaching them. In seconds, they were all out in the parking lot a hundred yards from the inferno. Five more fire trucks arrived.

"What the hell was that?" Vince said.

"Oh, I'm pretty sure that was meant for you and Mary Beth," Foster said.

"We're lucky," Mary Beth said. "If we hadn't been knocked over and trapped by that sheet of steel, we'd all be dead. It shielded us."

"No doubt," Foster said. He turned to the two FBI agents, "You guys are my heroes."

"Just doing our job," one of them responded. The other nodded.

"Alright, take these three to the van and get them out of here," Foster said, pointing to Bryce, Brown and Spaulding.

A dozen first responder EMTs arrived and began the search for any other survivors of the blast; it didn't look good.

Of the twenty-four boats at the Bay Sound Marina in Annapolis, fortunately, only six others had been occupied by their owners when the explosion happened. Of those six, four had miraculously survived—albeit barely—by being plunged into the water as their boats ignited and then exploded. The other two would never be found. Of the remaining eighteen boat owners and captains, eleven were still in the marina office building at the buffet and also managed to survive, although many were badly burned. The lucky rest of them were either out of town or merely out for dinner or drinks.

Despite the fire companies' best efforts and the million gallons of water and dispersal they sprayed, the fire continued to rage for three more days, fueled by the type two diesel JetA blanketing the surface of the water and the smoldering tons of teak and mahogany hulls of the vintage boats.

# CHAPTER THIRTY-NINE

**Ithaca**

Alina, Jason, Rose, Michael, Mary, Beth and Vince sat around the table in the back section of the Heights Cafe. The table setting was impeccable, the glasses spotless, the white tablecloth pristine, without a single wrinkle or ripple.

The server, dressed in sharply-creased black slacks, crisp white button-down shirt and black bowtie, circled the table, pouring the 1998 Veuve Clicquot La Grande Dame Champagne into each of their Waterford crystal champagne flutes.

Vince raised his glass, and they all followed.

"Here's a toast to my Mary Beth, my son, Michael, our good friend Jason, our survivor, Alina, and our artist Rose," he said. "We are all grateful to each other and for each other!"

They all clinked glasses and sipped.

"Before we order dinner," Vince said. "There's something I'd like to present Alina."

Mary Beth pulled a white envelope from her purse and handed it to Vince. He took it and placed it on the table in front of Alina, her name perfectly and artfully scribed on it by Rose across a crimson heart.

"This is for me?" She asked.

Vince winked at Mike, "It is for you, Alina."

She picked up the envelope and looked it over. "It's beautiful!" Alina said.

"Rose did the envelope," Michael said.

"Oh, thank you, Rose; it is so beautiful!"

"Open it!" Mary Beth said.

Alina carefully opened the envelope, withdrew its contents, examined it and set it down on the table,

"I don't understand. Another fake check? Is this a joke? Maybe a joke? Very funny!" she said with a light chuckle. She set it down on the table on top of the envelope.

"Ah, no. This one is real," Vince said.

"What?"

"Alina, while everything was going on, my dad filed the life insurance claim on your husband's life. This money is yours, and believe me, you've earned every penny," Mike said.

"Five million dollars?" Alina said in disbelief.

"Five million dollars, Alina," Vince confirmed.

Jason gasped.

Tears began to flow down her cheek. She put her face in her hands and wept. Mary Beth stood, walked behind her and hugged her. Little by little, Alina composed herself. She turned to Jason.

"Jason, I know what I want to do with this," she said. "I want to find and save all the girls that were taken by Russianbryde everywhere. Will you help me?"

"Of course I will, Alina."

"Rose, Mary Beth, I will need you too. Will you help me?"

They all nodded. Vince teared up.

# EPILOGUE

As part of his plea deal, Stan Reddinghouse, Russianbryde's former tech guy, worked with the FBI Regional Forensics Laboratory (RCFL) in Buffalo, New York. Chad Whitcomb, the Russianbryde algorithm specialist, was quick to agree to cooperate with the FBI to save his skin. They were able to track down four-hundred-sixty of Russianbryde's victims located in all fifty states of the U.S. within days, including Alina's friends from Tolyatti, Russia. Hundreds more would be located in the next few months.

With Vince's and Michael's guidance, Alina formed RB EMANCIPATION, LLC. and took up her place as CEO of the company. The company reached out to every victim of Russianbryde as soon as the victim was identified.

Jason Valor headed up the expansive team of the company's social workers; each Russianbryde survivor was assigned her own advocate. He worked network security along with the FBI in identifying and tracking illicit dating and "mail-order bride" sites.

Michael DeMarco was the lead counsel of the company, who assigned an attorney for each of the victims in all fifty states to guide them through any legal problems any of them faced. Rose oversaw the supply side of the company, providing survivors with clothes, shoes, coats, toiletries, and company debit cards with up-front funds of $2,000.00. She personally chose the styles and brands to be made available for any of the survivors and created a hand-painted catalog that was distributed to them. What was missing were prices for any of the items; they were all available for free.

Mary Beth surrendered her badge and gun but maintained daily contact with Assistant United States Attorney Sandy Mitchell, acting as a liaison between the victims, the FBI, and the

U. S. Attorney's office in following up on the countless crimes committed by Russianbryde and the many co-conspirators, enablers, and accomplices.

Alina's mother came and joined the company, acting as translator and general advisor to her daughter. She was also matron of honor at Alina's and Jason's wedding.

Rose introduced Michael to a world foreign to him: a world of artists, musicians, artisans and artists. She even convinced him to dress down and attend the Grass Roots festival in Trumansburg, New York—a miniature version of Woodstock. It was there he proposed to her, and she accepted.

Vince handled the company's purchase, financing and the new construction at what was left of the Bay Sound Marina, where they'd barely escaped with their lives. The remains were cleared out, and the company built its new headquarters there, replacing the destroyed piers and marina buildings with a swimming pool, tennis courts, and an area on the bay for kayaks and canoes. The office building and ten other cottages were built in the typical Chesapeake Tidewater style: simple lines, classic double-hung windows, vertical plank pine siding, cedar shingles, red brick fireplaces and chimneys. Each structure had a warm, vintage, welcoming, home-like presence and served as a safe shelter for victims of domestic violence, the first of many such shelters RB EMANCIPATION, LLC would establish in the years to come.

The Pennsylvania State Police spotted Big Red at a rest stop on the PA Turnpike. He was swiftly extradited to New York State and tried by District Attorney Gary Becker. There being only circumstantial evidence for a murder charge, he was charged with and convicted of manslaughter and criminally negligent homicide for the death of attorney Sean Adams. He was sentenced to ten years in state prison.

Ron Bryce Jr. at first refused any plea agreement despite overwhelming evidence against him. On the very morning of his trial, the jury seated, he succumbed and accepted the only plea bargain U. S. Attorney Sandy Mitchell was willing to offer: Guilty to conspiracy to commit murder, Guilty to heading a Continuing Criminal Enterprise, Guilty to wire fraud and money laundering. He threw himself on the mercy of the Court and was promptly sentenced to life in prison and ten million dollars in restitution to be paid to the victims of his crimes.

Bill Brown took a plea for twenty-five years to life with the possibility for parole in fifteen years—he would be ninety. Due to his expertise in sales, they put him in charge of the prison commissary.

## THE END

# CAST OF CHARACTERS

Vince DeMarco, Esq.
Michael DeMarco, Esq
Miranda, Paralegal
Mary Beth Allison, former FBI Special Agent
Jason Valor, Social worker
William J. Bill Brown-boat broker
Rick Anders-marina mechanic
Dr. Raymond Weber, MD
Alina Weber
Gary Becker, District Attorney
Sean Adams-Dr. Weber's attorney
Detective Cindy Maddison
Deputy John Wilson
Deputy Sam Irving, rookie
Tom Foster, FBI Special Agent
Sandy Mitchell, Assistant United States Attorney
Victor Bennington, prior owner Matthews 53' yacht
Virginia Bennington, Victor's wife
Randolph Everts, Esq. estate attorney
John Holdings, Owner 1930 Frank Anderson Yacht and Blue Crab Shack a/k/a Nikolai Volkov
John Miller, Owner 1929 75' Elco yacht
Randy Spaulding, Captain
Helen Stinson Weber, Dr. Weber's ex-wife.
Jennifer Britton, County Court Judge
Johnny Stevenson, Volkov's "target" in the sinking
MaryAnn Calais, collateral damage in the sinking
Naval Cadet Randy Emerson, collateral damage in the sinking, and MaryAnn's boyfriend
James Curtis, street name, Big Red, prior owner murder gun
Stan Reddinghouse, Russianbryde computer tech expert
Ron Bryce, Jr., Owner Russianbryde
Chad Whitcom, Russianbryde algorithm specialist

Made in United States
Troutdale, OR
11/03/2026

56845410R00215